ON THE

CRYSTALLOGRAPHY

OF CALCITE.

BY

J. R. Mc. D. IRBY.

BONN,
ADOLPH MARCUS.
1878.

TO

MY FATHER

AND TO

G. VOM RATH.

PREFACE.

The following study of calcite was occasioned by a prize-theme, "*a critical investigation of the skalenohedrons occurring on calcite*", which was set two years ago by the University of Bonn. Hessenberg had proposed the same problem to himself, and had collected notices of almost all known skalenhedrons. He had a series of cards printed, one for each form, arranged with appropriate spaces for the several angles and symbols. The filling ont of these cards had been only partially completed when he was suddenly snatched away. His remarks on the occurrence of the forms did not go back beyond Zippe. There was nothing to indicate that he had especially studied either Levy on Hauy. The number of skalenohedrons of which he had notices was 129. There were five angles indicated for each of them, those of the three edges, and the inclinations of the apical edges to the vertical. It was proposed to calculate the angles to seconds. About one fourth of them had been calculated. There was no indication of any system of classification which had occurred to him.

At Hessenberg's death his family committed these cards into the hands of his friend Prof. G. vom Rath, as the person most fitted to carry out his work.

Prof. vom Rath was however prevented by other occupations from immediately doing so; and, on my announcing my intention of working on the prize-theme, he did me the honor of handing these cards over to me.

At first I had no thought of studying any forms except the skalenohedros. The tables and accompanying remarks relating to them were a year ago so far advanced that they could be submitted for the prize, which I was so fortunate as to obtain; and I now determined to extend the ingestigation so as to include the remaining forms of calcite.

The whole of the work was performed while I was in Bonn, a student of Prof. vom Rath's. As teacher and friend I have everything to thank him for, but it is justice to him, as well as to myself, to say, that I have worked quite independently of him. Those acquainted with his writings will at once percive this.

The starting point of my investigation is the hypothesis of Hauy, modified by assuming as the *molecule integrante* the rhombohedron of 105° 5′ instead of that whose faces are equally inclined to base and prism. In adopting this hypothesis however I am perfectly in the dark as to what measure of defence or justification, is necessary, if any. I can call to my mind only two Germans works in which I find mention of Hauy. In the historical introduction to Quenstedts Mineralogy of course it could not have been omitted. Zippe in the introduction to his monograph on calcite (p. 1) says "the German crystallograpers very soon forsook the method of Hauy. Indeed one might say it has never become properly habilitated in Germany". I should not omit to name Scharff who in pursuit of analogies between crystals and plants has come to regard Hauy somewhat as the demon of crystallography — not a demon of that scientific and useful kind which figures in the mechanical theory of heat.

The impression produced is that Hauy, like au inconvenient topic, is ignored. And yet it seemes to me that all who fairly face the problem of the constitution of crystals must incline to the opinion, that he was in the main right. The theories of Bravais and Frankenheim are in no essential particular different from that of Hauy. If in their *assemblages* we consider the shape of that portion properly belonging to each point we arrive at once at the *molecules integrantes.*

The constitution of the crystals being that described Millers system of axes is the only natural one. In selecting these for calcite, however, I do not mean to advocate the entire replacement of the hexagonal axes by them. As Bravais has shown, there is a rhombohedral and also a hexagonal system.

The method I have employed in investigating the development of the crystal is, so far as I know, new, though it may be that Sadebeck in his "Krystalltektonik" has treated crystals in an analogous manner.

In the tables I hope few errors will be found. I have taken great care both in calculating and in correcting the printer. The angles have been calculated to seconds but this is rather an over refinement; for we are not sure of the fundamental angle within a minute[1]). Many other details which might have been added to the tables have been omitted for fear of incumbering them. It might perhaps have been well to have added a separate table with the symbols of the forms according to Weiss and as they are given in the *Zeitschrift für Krystallographie.* But whoever regards the intercepts on Millers axes as expressing the constitution of the faces will have but little sympathy with these nomenclatures and would only add them in order to make the work more accessible to others. This I considered practically to have been attained by the use of Naumann's symbols. Weiss' intercepts may at a glance be changed to m R n, and there is no paper on calcite or book on mineralogy, in which the indices of the *Zeitschrift* have been used unaccompanied by other symbols. The symbols of Levy may at once be written from Miller's indices. $d^{1/h}$ $d^{1/k}$ $b^{1/l}$ =

1) Where no previous calculation of an angle was at hand for comparison, it has been calculated twice. If my recalculations of Zippe's angles agreed with his to the minute (they are only carried so far), I have not repeated them.

$(h\,k\,\bar{l})$. $b^{\frac{h}{k}} = (h\,k\,o)$. $d^{\frac{h}{k}} = (h\,o\,\bar{k})$. $a^{\frac{h}{k}} = (h\,k\,k)$. $e^{\frac{h}{k}} = (h\,\bar{k}\,\bar{k})$. $e_{\frac{h}{\bar{k}}} = (h\,k\,\bar{k})$.

The critical portions of the work have not always been pleasant especially where determinations of vom Rath and Hessenberg had to be called in question. The uniform and genial kindness which I have experienced from the former during my term of study in Bonn has bound me to him by the firmest ties. With Hessenberg I had not the honor of an acquaintance. He died before I came to Europe. But his signal achievements without teacher or early training, and his near relations to Prof. vom Rath, and also to my self, as the continuer of his work might well inspire me with the highest reverence for him. I am sure that Prof. vom Rath is, and that Hessenberg would have been, convinced that my one object has been to arrive at the truth.

My thanks for many kindnesses are due to Prof. C. Klein, now at Göttingen, formerly of Heidelberg, my first teacher in crystallography.

J. R. Mc. D. IRBY.

ON THE CRYSTALLOGRAPHY OF CALCITE.

LITERATURE.

I. Of the faces.

Hauy. Traité de Mineralogie. 2me Ed. Paris 1822[1]).

Bournon, Comte de. Traité complet de la chaux carbonatée et de l'Arragonite. Londres 1808.

Monteiro. Journal des Mines. 1813. II. p. 194.

Monteiro. Annales des Mines. 1820. V. p. 3.

Wackernagel. Kastner's Archiv für die gesammte Naturlehre. Bd. IX. Kritik über Bournon.

Weiss. Grundzüge der Theorie der Sechs-und-sechs-Kantner und der Drei-und-drei-Kantner. Berl. Akad. 1822—1823. p. 217.

Weiss. Fortsetzung der Abhandlung, „Grundzüge der Theorie u. s. w.", insbesondere über die von Herrn Levy neu bestimmten Kalkspath-Flächen. Berliner Akad. 1840. p. 137.

Weiss. Neue Bestimmung einer Rhomboëder-Fläche am Kalkspath. Berl. Akad. 1836. p. 207.

Weiss. Ueber herzförmig genannte Zwillinge von Kalkspath und gewisse Analogien von Quarz. Berliner Akad. 1829.

1) Hauy has written many smaller notices on Calcite in journals &c. but this edition of his Mineralogy published just before his death may be taken to cover all: that is all that does not properly find a place in his Traité de Crystallographie.

Naumann. Pogg. Ann. 1828. Bd. XIV. 235.

Levy. Description d'une collection de Mineraux formée par M. Heuland. Londres 1837.

Hausmann. Handbuch der Mineralogie. 2. Aufl. 1847.

Breithaupt. Vollständiges Handbuch der Mineralogie. 1836—1847.

Zippe. Uebersicht der Krystallgestalten des rhomboëdrischen Kalkhaloides. Denkschriften der Wiener Akademie. Bd. III. 1851.

Hochstetter. Das Krystallsystem des rhomboëdrischen Kalkhaloides; seine Deduction und Projection nebst einer Vergleichung mit der Entwickelung des Tesseral-Systems in rhomboëdrischer Stellung. Denkschriften der Wiener Akademie. Bd. VI. 1854.

Sella. Studi sulla mineralogia Sarda. Mem. d. R. Acad. d. scienzi di Torino. Serie II. Tom. XVII. 1856.

Sella. Quadro delle Forme Cristalline del Argento Rosso, del Quarzo et del Calcare. Torino. 1856.

Hessenberg. Mineralogische Notizen in den „Abhand. der Senckenbergischen naturforschenden Gesellschaft". Frankfurt am Main [1]).

Bd. III. Kalkspath von Rossie St. Lawrence Co. N.-Y.
„ „ unbekanntem Fundort.
„ „ Andreasberg.
„ „ Maderaner Thal.
Bd. IV. „ „ Bleiberg.
„ „ Maderaner Thal[2]).
„ „ Ahrnthal in Tyrol.
„ „ Matlock.
„ „ Andreasberg.
Bd. VI. „ „ Island.
Bd. VII. „ „ Lake Superior.

---

1) I have always in referring to these given the volume of the „Abhandlungen", in which they are to be found. Hessenberg's designation was ingeniously confusing. Each paper bears three numbers, that of the whole number of papers that of the new series and that of the continuation.

2) Discovered later to be dolomite from Binnenthal.

Bd. VII. Kalkspath von Agaëte auf Gran Canaria.
Bd. VIII. „ „ Bleiberg.
„ vom Rödiefiord auf Island.
Bd. X. „ „ „ „ „
„ von Andreasberg.

G. vom Rath. Mineralogische Mittheilungen in Pogg. Ann. The following numbers are on Calcite.
No. 19 in Vol. 132.
„ 20 „ „ 132.
„ 24 „ „ 135.
„ 49 „ Ergänzungs-Band 1871.
„ 76 „ Vol. 152.
„ 83 „ „ 155.
„ 90 „ „ 158.

G. vom Rath. Mineralog. Mittheilung.: neue Folge, in Groth's Zeitschrift für Krystallographie. No. 5. Vol. 1.

Descloizeaux. Manuel de Mineralogie. Tome I et Atlas. Paris 1862. Tome II prem. Part. 1874.

Kokscharow. Ueber d. russ. Kalkspath. St. Petersburg. Acad. VIIIth Series. Vol. XXII[1]).

Peters, C. F. Ueber Kalzit u. die rhomboëdrischen Karbonspathe im Allgem. N. Jahrbuch f. Mineralogie. 1861.

Scharff, F. N. Jahrbuch für Mineralogie for 1860 u. 1862.

Scharff, F. Ueber den inneren Zusammenhang der Krystallgestalten des Kalkspaths. Frankfurt a. M. 1876.

Websky. Tschermak Min. Mittheil. 1872. p. 63.

Zepharovich. Mineral. Mitth. Sitzungsberichte d. Wiener Akademie. Vol. 54. p. 273.

Zepharovich. Mineralog. Lexicon für Oesterreich. Wien Bd. I. 1859. Bd. II. 1873.

Dana. Mineralogy. 5th Edition.

Schnorr. Studien über d. Mineralien v. Zwickau. Zwickau 1874.

---

1) The same paper with a few unessential additions is contained in the seventh volume of his „Materialien zur Mineralogie Russlands".

Frenzel. Min. Lexicon für das Königreich Sachsen. Leipzig 1874.
Wimmer. Kalkspath von Andreasberg. Ber. u. d. dritte gen. Ver. d. Clausthal. naturwiss. Vereins. Maja 1854.

II. Of Twins.

The principal sources of our knowledge of Twins of calcite are Sella's „Studi etc.“ and vom Rath's Mineral. Mitth. Nr. 20. Incidental discriptions of them will be found in a large number of the above works. In addition to these the following may be mentioned.

F. v. Kobell. Kastner Archiv f. die ges. Naturlehre. XIII.
F. v. Kobell. Journal für prakt. Chemie. XVIII.
Scheerer. Pogg. Ann. Vol. 65.
Scharff. N. Jahrbuch f. Mineralogie. 1870.
G. Rose. Ueber die im Kalkspath vorkommenden hohlen Canäle. Berl. Abhand. 1868.
G. vom Rath. Mineral. Mitth. neue Folge. No. 9. Groth's Zeitschrift. Vol. II.

III. Of microscopical Crystals of Calcite.

G. Rose. Ueber die Bildung des Kalkspaths und Arragonits. Pogg. Ann. XLII. 1837.
G. Rose. Ueber die heteromorphen Zustände der kohlensauren Kalkerde. Berl. Abhand. 1856—1858. Monatsbericht. 1860.
P. Harting. Etude microscopique des precipités et de leurs metamorphoses. Bulletin des sciences physiques et naturelles de Neerlande. 1840.
P. Harting, in the Tydschrift voor Natuurlyke Geschiedenis und Physiologie. 1843.
Link. Bildung der festen Körper. Berlin 1841.
Vogelsang. Die Krystalliten. Bonn 1875.
H. Credner. Journal für praktische Chemie 2. (2). p. 390.

The above list contains the essential sources from which our crystallographical knowledge of calcite is derived. In these are contained notices of about 200 faces, twins according to 6 laws, and more than 1000 combinations.

Among the faces are many which require very high numbers to express them, these varying greatly according to the system of axes to which the faces are referred. In some cases these numbers are of four figures, and in not a few of three. It is evident that the law of simple rational indices as usually stated would have to be considerably modified if these faces be included under it.

What is therefore proposed, is as far as possible to ascertain which of these faces and kinds of twins have been correctly determined and to find the most natural and simple system of axes to which to refer them. Also to ascertain and exhibit the great general features of the crystallogenic development of calcite.

Some remarks are necessary on certain of the above works.

Bournon's treatise is interesting historically but of little value. Wakkernagel has criticised it severely and Zippe fully confirms his judgement. It seems to be so full of inaccuracies as to be quite unreliable. I have, therefore, not used it at all and have given as doubtful all faces in Zippe which rest solely on its authority.

Hessenberg and vom Rath have been the principal workers on calcite since the publication Zippe's monograph. It is especially in their papers that the extremely complicated forms have been described. I learned first from Prof. Groth of Strassburg, that Hessenberg had made all his measurements with a goniometer without telescope and he suggested to me to repeat them. His collection having been purchased by the University of Halle, I was enabled through the kindness of Prof. von Fritsch, the director of the mineralogical museum there, to submit the most interesting of

Hessenberg's combinations to a careful study. My results are contained in Appendix I. They differ very widely from those of Hessenberg but I think they will be generally found to be correct. My best thanks are also due to Dr. C. Bodewig, of Köln, for placing at my disposal his goniometer by Fuess and in fact his whole laboratory.

I. Of the constitution of crystals of calcite.

In order to ascertain the proper system of axes to which to refer the faces of calcite we will study the composition of one of its crystals as exhibited by the cleavage.

I think it may be fairly assumed that, accidental irregularities being excepted, all cleavage planes of calcite are similar.

This being so, suppose one of its crystals cleaved up to that limit at which the rhombohedrons obtained can not be further cleaved. It is evident that the particles so obtained have a certain distinct existence in the uncleaved crystal; for the forces which unite them one to another are different from those which bind the particles within them together.

Our knowledge of a crystal of calcite would therefore be complete, if we knew

1) the constitution of the ultimate rhombohedrons;
2) the manner in which they are arranged in the crystal;
3) how they are limited.

All we know of the composition of the ultimate rhombohedrons is, that such is the space they occupy in the crystal and that chemically they are Ca CO^3.

Of modes of arrangement only one is possible, if they be brought fully to fill a space, and that is when they are similarly placed and in contact with one another. Such is the arrangement exhibited by the cleavage.

The constitution of the faces may not be directly observed and we can only show what it probably is.

Assuming the homogeneousness of the crystal two hypotheses as to their nature are possible. We may either suppose, as Hauy did, a face to result from the piling up of elemental rhombohedrons so that it is really the limit of a regularly broken surface: or, we may suppose the elemental rhombohedrons capable of division and the interstices in a face of the above character to be filled up by fractions of rhombohedrons so that every face would be continuous in the same sense as the cleavage planes.

If we take the latter hypothesis, we shall not only have to suppose the fractions of rhombohedrons to exist in the completed face, but also in every stage of its development: and, since the known cleavage is common to every part of a crystal, we should have to suppose the fractions deposited in one instant continually being completed in the next to whole rhombohedrons, such as are obtained by cleavage. If crystals were built up in this manner it would certainly be expected that there would be a cleavage parallel to each face. In addition to this, in the case of a crystal of many faces in the act of growth we should have to suppose as many different kinds of fractional rhombohedrons being deposited as there were different faces on the crystal. We should have further to assume that the rhombohedrons were capable of being divided only as would be required by the law of rational indices; and, that they were more easily divided when the mode of division could be expressed by simple indices. It would continually be necessary to add still other suppositions as to the character of the fractional rhombohedrons in order to account for each accident in the condition of a face, for this hypothesis allows of no variations in the grouping of the elemental rhombohedrons. In short it will be seen, that it would explain exactly what we assume and nothing else. The hypothesis of Hauy alone remains. The great difficulty with it is to understand by what possible means some of the complicated faces known could be formed. The following may in same measure remove this.

From the phenomena of supersaturation of solutions and sudden deposition of the dissolved substance in a crystalline condition we may, on the principle of continuity, infer, that the ordinary growth of crystals proceeds in a similar manner, but on a smaller scale. That is to say the process consists of a series of surchargings of the solution and dischargings of crystallising body.

From the difference of constitution of the several external parts of a crystal, as for example corners and edges, we may infer that, in the act of growth these different parts receive particles from the solution with various degrees of ease. That is to say, crystals grow at differrent rates in different directions. To illustrate by an extreme case, needle shaped crystals seem to be capable of growing almost only in one direction. Such crystals having, however, some thickness and that a varying, it is plain that one direction of growth only preponderates greatly over the others; does not entirely supersede them. On the principle of continuity, we should expect this to be true in a modified degree of all crystals.

Imagine now a crystallising solution and AB the face of a growing crystal. As an example we may take a rhombohedron of calcite. Let the plane of the paper be one of its faces; AB and Ab the traces of the other faces upon it. Suppose a portion of $Ca\,CO^3$ discharged on the corner A. This would arrange itself, in virtue of the crystallogenic forces, as one or a group of several elemental rhombohedrons. Suppose it to be the former. A new corner similar to A would thus be produced. This would be prepared to repeat the same process as soon as the solution about it were again sufficiently concentrated. This would continue to reoccur at constant internals, as long as the rate of evaporation of the solvent remained unchanged. In the meantime the face cd has come into existence and is also prepared to grow.

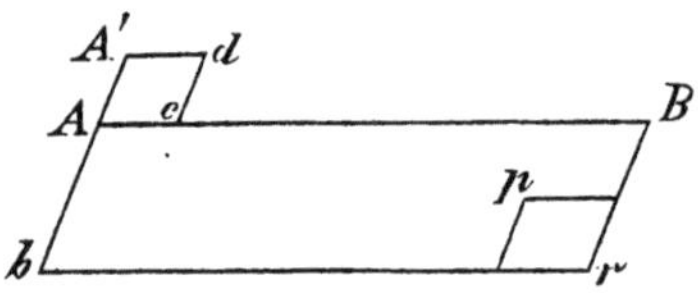

The circumstances of A'd and cd are however entirely different. If the faces meeting at A' and c be both attracting particles from the solution, as is probably the case, the resultant attraction in once case is directed outwards from A' and in the other lies within angle at c. Its intensity in each case would depend not only upon the specific attracting power of the surfaces concerned, but also upon the angle under which they meet. This difference of intensity and relative direction of the forces at A' and c would cause different rates of growth at those points. As soon AA'dc was deposited the face cd would begin to grow in the direction AB at a certain rate. After a certain interval another rhombohedron would be placed on Ad, and a new corner, precisely similar to that at d, would be produced, and a new course of particles would grow out in the direction AB. Since the conditions of the two are similar, this course would grow at the same rate as the first course and would remain at a constant distance behind it. Thc same would be true of all subsequent courses with respect to the courses immediately preceeding them, so long as the conditions of constant growth were fulfilled. If the thickness of the courses and the intervals of their extremities remain constant, it easy to see that a straight line is the locus of the latter.

A similar procedure to the above would occur in the face whose trace is Ab. It remains now to see what occurs in face whose trace is AB, on which the particle AA'dc was originally deposited, besides what we have just seen in the extreme row AB and the similar one in the plate whose trace is Ab.

We may use the lower right-hand corner of the same figure to represent this. Suppose this to be the face AB, seen from above, and rp the particle AA'dc. Two cases are possible: the growth in the directions rb and rB may proceed simultaneously or that in one direction may preceed the other. Suppose the latter more general case and that the growth in rb first begins. The first course of particles would be laid on under singular circumstances. It would proceed along the edge rb and would have no course on either side against which to be laid. The first particle in the direction rB would form the beginning of a new course in the direction rb, which could grow in a precisely similar manner to the first course in AB. The only difference between the two cases is the presence in this latter of the face upon which the whole growing takes place. Its action would be constant, however, on all the particles and would cause no variations in the rate of growing, though it might change its absolute value.

As long as the row of particles on the edge that first began kept a head of the second row, the state of things first described as occurring in the plane bAB would be maintained, succesive layers of particles would be laid on and as long as their rates of growth remained constant their extremities would be found in a straight line. The second particle deposited in the direction AA′ would begin a second similar layer, a certain distance behind the first, and so would the third and following particles. The straight lines in which the limiting particles of each layer lie would all be contained in one plane so long as the conditions of growth remained constant: for then the thickness of the layers and their distances from one another would also remain constant. The extremities of the first rows in each layer would probably not be in this plane, because the singular conditions of their growth would probably make their rate of growing different from that of the others.

The intercepts of this plane on the three lines of growth would at any instant be respectively equal to the velocities in them multiplied by the time elapsed from the beginning. The time being the same in all three they would be proportional to the volocities. The velocities being expressible by the number of rhombohedrons deposited in the unit of time, they are necessarily expressible by whole numbers. The intercepts on these lines will therefore be rational whole numbers, if we take the thicknesses of the elemental rhombohedron in the given directions as the units.

In the foregoing discussion two limitations have been made. It was assumed that the growth from the corner r in the two directions r B and r b did not begin simultaneously, and also that the first row of particles laid down along the edge r b did not grow slower than the second. We need not consider what would occur when these restrictions were removed. Our object was to show how in virtue of forces, ascertained to be present in the act of crystallisation, the elemental particles of a crystal could be arranged so that their limiting surfaces would be planes, hoping thereby to remove some difficulties in the way of accepting Hauy's theory of the constitution of crystals. To pursue the subject further into details would be useless without a special knowledge of the forces concerned.

It will be observed that the essential element above considered is time. This is probably only one of a number affecting a growing crystal. The final result would be due to their joint action and it may be that in ordinary cases this one would produce but little effect. We should expect it, however, to be asserting itself distinctly in extreme cases, that is, in those cases in which the time was either very long or very short. When crystallisation takes place very quickly we know that the resulting crystals are generally imperfect. The following may help us to understand what occurs when a great deal of time is allowed. Suppose the solvent to be evaporating very slowly and the solution to be in extreme stillness. Under these circumstanes the crystallizing body would be supplied for arrangement much more slowly than the forces of crystallisation were capable of arranging it, and an attraction might be exercised by centres of stronger growth, which would cause streams of particles to flow to them. They would thus receive more particles than centres of weaker growth. In such a case the rates of growth would be different; and they might have any ratio to each other. We should rather expect complicated faces than otherwise; for there is no apparent reason why these ratios should be simple numbers. The form — $^{31}/_{20}$ R. $^{67}/_{31}$ in the combinations from Elba, described by vom Rath (Pogg. Ann. Bd. 158, p. 414) might well have been produced in this way, since it has no connection with the accompanying faces by which a zonal or other known development might be supposed.

The hypothesis of Hauy may therefore be assumed. It is not in conflict with any known facts, it explains many, and it is probable that it will explain all when the crystallogenic forces have more completely investigated.

As the axes to which faces are to be referred we shall take edges of the primitive; as the origin its apex. We shall now show how a face with any given intercepts may be built up; and, how the constitution of a face may at once be inferred from the same.

Let a, b, c be intercepts of the face, deprived of all common factors and expressed in terms of the unit-length = edge of elemental rhombohedron. Its equation will be

$$\frac{x}{a} + \frac{y}{b} + \frac{z}{c} = 1.$$

It is evident, that for every point of this surface, whose coordinates are whole numbers, an arrangement of elemental rhombohedrons may be made so that a corner of one of them coincides with this point; and, that such points are the only points of the surface, which may be occupied by corners of elemental rhombohedrons arranged in the required manner. The distribution of these points expresses the physical constitution of the surface. The general values of x, y, z of these points are as follows.

In the case that no two of the intercepts have a common factor, they are of the form ma, nb, pc, where m, n, p are any whole numbers. The set of values corresponding to one point must satisfy the condition $m + n + p = 1$.

In the case that any two of the intercepts have a common factor certain values have to be interpolated between those just given. Suppose a and b to be two such and q the common factor. Then all values of x and y of the form

$\frac{ma}{q}$ and $\frac{nb}{q}$ will also belong to such points, m and n being as before any whole numbers. Similarly for factors common to b and c, or c and a.

We may express these results geometrically in the following manner.

Let one picture to ones self how the elemental rhombohedrons must be arranged in order to build up the basal face (111). An exactly corresponding arrangement would be that of cubes to build up an octohedron.

In order to build up the above face $\left(\frac{1}{a}\frac{1}{b}\frac{1}{c}\right)$, take, instead of simple elemental rhombohedrons corresponding to (111), a compound rhombohedron containing a b c elemental rhombohedrons, and whose edges are respectively equal to a, b, c. Arrange these precisely as the elemental rhombohedrons were arranged for (111) and so that all similar edges are parallel. The limit of such an arrangement is the given face. This is the complete arrangement for the case that no two of the intercepts have a common factor. If this is not the case something must be added to the above. The above arrangement cut by a plane parallel to x y would give a step-formed line, the parts of which parallel to the x-axis were a units long and parallel tothe y-axis b units long. If a and b have a common factor q these steps must by interpolation be reduced to the lengths $\frac{a}{q}$ and $\frac{b}{q}$ in the given directions. It will be seen that this may be effected for the general arrangement by supplemental rhombohedrons whose edges are $\frac{a}{q}$, $\frac{b}{q}$, c. In case that two or more pairs of the intercepts have a common factor the interpolation for each pair should be made in exactly the same manner. It will be found however in the final result, that the different supplemental rhombohedrons used have a certain number of elemental rhombohedrons common.

It will be seen from the above that faces whose intercepts, taken in pairs, have common factors, have a different complete arrangement from those which have not.

We take the axes of Miller therefore as the rational system for calcite. The necessity of referring all faces to these axes becomes especially apparent in the case of faces whose intercepts are expressed by certain prime numbers. In these cases in order to judge if the intercepts be not too great we must also go behind the indices and use the intercepts themselves. For example a form whose intercepts were 13, 17, 19 would be quite admissible. Its indices would howerer be 323, 247, 241 and its symbol according to Naumann $^{70}/_{611}$ R. $^{282}/_{35}$. The reverse is also true. Symbols, quite admissible according to other axes might be assigned to faces, which, being expressed according to Miller's axes, would be perfectly inadmissible. A form $^{19}/_{5}$ R. $^{19}/_{10}$ furnishes an example of this. If the magnitude of m and n in mRn were alone to be considered, this form would be admissible. The indices, however, according to Miller, would be 1373, $\overline{280}$, $\overline{793}$ and since there is no factor common to any two of these, the intercepts would be 280×793, $\overline{1373\times793}$, $\overline{1373\times280}$.

II. Of the development of the Crystals.

a) The primitive form.

The following observations refer to this.

Harting[1]) found that the precipitate produced by mixing concentrated solutions of a calcic salt and an alkaline carbonate is at first perfectly colloidal and homogeneous. Almost as soon as it is formed, however, an extremely fine granulation appears, which increases in size until the granules become somewhat less than 20 mm. in diameter and assume a certain structure. These sometimes remain

1) P. Harting, Recherches de morphologie synthetique sur la production artificielle de quelques formations calcaires organiques. Amsterdam 1872. p. 5.

permanent, sometimes pass into rhombohedrons. The number of the latter and also the rapidity with which these changes follow one another increases with the temperature.

Link made nearly similar observations.

Gustav Rose [1]) has studied the same subject. He found that if calcic carbonate is precipitated at ordinary temperatures and allowed to stand it beomes coarsely granular and on examination under the microscope is found to consist „*wholly of sharply limited, transparent, and perfectly distinct rhombohedrons*"; which, he adds, like those about to be described, are evidently the principal rhombohedron. He also found, that when a solution of $Ca\,CO^3$ in H^2O+CO^2 was allowed slowly to evaporate at ordinary temperatnres the crystalline crust obtained consisted of rhombohedrons, commonly with the basal face. Certain of these were long enough to be measured with the reflecting goniometer. The rhombohedron was that of 105° 5'. He also found [2]) that natural earthy deposits of calcite are generally composed „*entirely of little rhombohedrons, which, as a rule, are the principal rhombohedron*". He obtained the same form, when he precipitated $Ca\,CO^3$ in the form of arragonite and then allowed it to transform itsself into calcite. The calcite contained in the pseudomorphs of calcite after arragonite from Herrengrund, Hungary, consists of rhombohedrons, sometimes partially modified and sometimes completely changed into other forms. Calcite obtained by dropping pieces of an alkaline carbonate into fused calcic chloride and afterwards removing the soluble substances with cold water also took the form of the cleavage rhombohedron.

Credner has made observations on artficial crystals of calcite obtained from a solution of $Ca\,CO^3$ in H^2O+CO^2. He obtained rhombohedrons which he determined by measurement to be that of 105° 5'. Occasionally the summits and the lateral corners were modified. When foreign sub-

---

1) G. Rose. Pogg. Ann. XLII, 1837.

2) G. Rose. Berliner Akad. 1856. p. 12.

stances, such as sodic and potassic silicates, were introduced into the solution, many of the crystals assumed a greater variety of form. The figure of the principal rhombohedron was, however, seldom widely departed from, but this was often modified at the corners and edges; and in some cases rhombohedrons, whose faces were replaced by pyramidal elevations, formed of two skalenohedrons, were obtained.

Vogelsang (Krystalliten, p. 87, u. Taf. XI. Fig. 1) has observed, under a magnifying power of 800, the polysynthetic character of the rhombohedrons resulting from the transformation of globules, such as described in the above experiments of Harting.

These observations may be divided into two classes: those in which the changes of $Ca\,CO^3$ in states antecedent and leading up to the crystalline were observed, and those in which the crystals were observed after their beginning. The forms assumed in those cases where the actual formation of crystal could be observed were almost without exception the rhombohedron of 105° 5′. In the other cases this form was taken by the bulk of the crystals but a greater or smaller number were modified. From a pure solution of $Ca\,CO^3$ in H^2O+CO^2 a much larger proportion of principal rhombohedrons was obtained than when foreign substances were introduced. In the former case the modifications were also simpler; they consisted almost only of planes truncating the corners of the principal rhombohedron. In the latter case in addition to these, the edge zone of the primitive was sometimes more or less developed. The combination observed were (100) (100), (100) (310) $(20\bar{1})$, (310) $(20\bar{1})$ and some others. The skalenohedrons could not be measured but they had the appearance of those named. In the rare combinations in which (100) was not actually present a rhombohedron-like form was observed. I think the presence of so large a number of primitive rhombohedrons and the gradual departure of the other forms from it indicate that the latter were also developed from it. These observations were made on calcite formed in such a variety of ways that one may confidently expect to find this *beginning of the crystals with the rhombohedron of* 105° 5′ *to be a general crystallogenic law of calcite.* This is exactly what would be expected according to the hypothesis of Hauy and we may, as he did, with all propriety call the rhombohedron of 105° 5′, *the primitive.*

If we consider the mode of growth of the primitive crystals, it consists simply of increase of size without change of form. Their crystallogenic increment would, therefore, also have the form of the primitive rhombohedron. Accordings to Hauy's hypothesis its form is the same for all crystals in all stages of their development.

b) The further development.

We have seen that probably all crystals of calcite have the rhombohedron of 105° 5′ for their primitive form: and also that the beginning of the further development is the apparance of faces truncating its corners and edges, and occasionally of two skalenohedrons in the edge zone of the primitive. We shall now attempt to trace this development further and to discover its principal features. A knowledge of the comparative frequency of the several faces and combinations is necessary for this purpose.

The only source of exact information on these points is Levy's *Description d'une collection de mineraux formée par M. Heuland.* This, as the title implies, is a crystallographical catalogue of a whole collection of minerals. It is the only one we possess. It contained 346 specemens of calcite with determinate forms, representing 198 combinations. I have counted the number of specimens on which each face occurs and also the number of combinations which are formed by certain sets of faces. The following table contains the result of the count for the faces. With reference to these figures, and also those hereafter to be given relating to the combinations, it is to be remarked that they are probably vitiated by the following causes. A collection is in general a selected set of specimens, and consequently, the commoner varieties are too smally represented. We should therefore expect the figures referring to the more frequent faces and combinations to be proportionately too low. Further, in this particular collection three localities were very much more largely represented than any others. These were Derbyshire, Andreasberg and Alston Moor. Out of the 346 specimens 96 were from the first, 90 from the second and 31 from the third. The other localities were represented by much smaller numbers of specimens.

Face.	No. of specimens on which it occurred.
$(2\bar{1}\bar{1})$, ∞R	232
(110), $-\frac{1}{2}R$	159
$(20\bar{1})$, $R3$	110
$(1\bar{1}\bar{1})$, $-2R$	88
(111), $0R$	78
(100), R	45
$(3\bar{1}\bar{1})$, $4R$	43
$(1\bar{1}\bar{1})$, $\infty P2$	41
(310), $\frac{1}{4}R3$	30 (24 from Derbyshire)
$(4\bar{5}\bar{5})$, $\frac{3}{2}R$	20 (12 from Andreasberg)
$(21\bar{2})$, $-2R2$	19 (11 from Alston Moor)
$(40\bar{3})$, $R7$	19
$(30\bar{2})$, $R5$	16
$(9\bar{4}\bar{4})$, $13R$	16.

The remaining faces occurred on a still more limited number of specimens. In the table of occurrence detailed remarks will be found upon them.

The extremely limited number of specimens, only about $\frac{1}{7}$, on which the primitive occurs, is, in the first place, to be noticed. We infer from this, that the modifications, which we have seen beginning on microscopical crystals, advance on about $\frac{6}{7}$ of those which attain to macroscopical proportions so far as to leave no trace of the form from which they proceeded. As a simple form (100) is extremely rare, with the exception of one locality. Levy records only one such. Descloizeaux mentions, however, that it is the most common of all combinations on Iceland Spar. From the purity and size of these crystals it may be inferred that they were formed in solutions very free from foreign substances and also from disturbances. This would render it probable, that when left entirely to itself calcite would develop only in this form. This is to a certain extent supported by the observations of Credner on artificial crystals. He found the rhombohedron (100) to predominate much more largely over the other forms when the crystals were obtained from pure solutions than when foreign substances were introduced. When $Ca\,CO^3$ in $H^2O + CO^2$ alone was taken the only additional forms observed were (111) oR and a plane truncating the lateral corners. On crystals obtained in a similar manner, but under perhaps less constant conditions, (Credner preserved his solutions from changes of temperature in a vault.) Gustav Rose found (111) commonly present. It might be, that under still more constant conditions and with greater masses of solution, these two faces would also not be formed. Credner observed that the further growth of his crystals changed them to rounded irregular forms, which indicates that the conditions of growth were not perfectly constant.

In combination the primitive seems to be pretty generally distributed in the different localities. Its absence was most conspicuous, however, on the Andreasberg crystals. It was only found on 3 out of 90 specimens from that locality. It seems on the contrary to be very common on the specimens from the Departement de l'Isere. Of these 9 out of 15 had it. One third of the whole number of specimens on which it was found were from Derbyshire.

(211), ∞ **R** occurred on $^2/_3$ of the specimens. This as Zippe has also remarked, is the most frequent of all forms. Its general office will, I think, be found to be the modification of the lateral corners of the primitive. The planes which Credner observed in this position are described as prisms and also as steep negative rhombohedrons. The latter were probably $(9\bar{5}\bar{5})$, — 14 R, the most frequent form of that kind. The two forms seem be capable of replacing one another, for they seldom occur together. I find $(9\bar{5}\bar{5})$ noticed 25 times but only twice in comb with $(2\bar{1}\bar{1})$.

(110), — $^1/_2$ R occured on somewhat less than one half of the specimens. Its general function is undoubtedly to truncate the apical edges of the primitive. It is found in this position on microscopic crystals. With $(2\bar{1}\bar{1})$ it forms perhaps the most frequent of all combinations on calcite. It is almost invariably striated parallel to that edge of the primitive which it truncates; indicating thereby that it is really a succession of these edges lying in one plane. In this office it seems to be both replaced and supplemented by the faces which bevel this edge: the latter more frequently. In Levy such faces occurred 27 times without (110), 18 times with it.

$(20\bar{1})$, R 3 was on about $^1/_3$ of the specimens. Its function is to modify the lateral edges of the primitive and as such has been observed microscopically. According to Gustav Rose (Pogg. Ann. Bd. 91, p. 147) this form is that most commonly taken by the calcite formed by the gradual transformation of the crystals of Arragonite from Herrengrund, Offenbanya and Girgenti. It seems to represent a very stable kind of development, as is shown by its great prominence on almost all very large crystals of calcite with the exception of those from Iceland. Even on these it is present in all observed combinations other than (100) and (100), (111).

$(1\bar{1}\bar{1})$, — 2 R occurred on about $^1/_4$ of the specimens. There is no indication that it plays any one prominent part. It may sometimes replace $(2\bar{1}\bar{1})$ at the corners; but it occurs oftener with than without it. Levy found the two together on 51 specimens, $(1\bar{1}\bar{1})$ occurred otherwise on 34. It is perhaps very easily formed on account of its simple relation to the primitive; the edges of the first being truncated by the second. The potency of this relation is indicated by the fact that fourteen other faces have it. Very remarkable is the occurrence of this form at Fontainebleau in crystals which consist of $^2/_3$ sand, the remaining $^1/_3$ being $CaCO^3$. A similar occurrence of R 3, $(20\bar{1})$ has been observed near Heidelberg. One should expect a form with such very determined crystallogenic forces to have asserted itself on the largest crystals. I am not aware, however, that it has been so observed, unless in a secondary capacity, as truncating the X-Edges of $(20\bar{1})$. As regards localities, it was well distributed in Levy.

(111), 0 R occurred on something under $^1/_3$ of the specimens. It truncates the apices of the primitive and is after it the most frequent form observed on microscopical crystals. Credner remarks that in such cases it always presents a darkened appearance. Gustav Rose (Pogg. Ann. Bd. 42, 1837) describes the crystals obtained by the gradual escape of CO^2 from the solution of $CaCO^3$ in $H^2O + CO^2$ as commonly having this form and (100). Most marked is the partiality of it occurence in Levy. Of 96 specimens from Derbyshere not a single one had it and of 31 from Alston Moor only two had it. On the other hand 58 out of 74 specimens on which it did occur were from Andreasberg. The total number of specimens from the letter place was 90. Its mode of occurrence at Andreasberg is also singular. It usually forms a white opaque layer terminating the crystal, and quite different from the remaining portion of its substance.

$(3\bar{1}\bar{1})$, 4 R occurred on about $^1/_3$ of the specimens. There is nothing to show that it has any special functions. On 29 out of 43 specimens it occured with $(2\bar{1}\bar{1})$. This suggests that its principal office is to modify the edge $2\bar{1}\bar{1}$: 100. It has the lowest indices of any face lying betwen these two. $(9\bar{4}\bar{4})$ 13 R seems generally to perform the same duty, for it occurs 22 out of 30 times with $(2\bar{1}\bar{1})$. Its geometrical relation to $(1\bar{1}\bar{1})$, — 2 R seems to have little efficiency for its formation, since the two occur together only 6 times in 43. Its geometrical relation to $(20\bar{1})$ might however be efficient, for the two occur together 25 times out of 43. It occurs generally in combinations of more than the average number of forms, which would seem to indicate, that it was a face easily produced in the course of development and under various conditions. It occurs at a considerable number of localities (see Zippe p. 29) as a simple form. In combination it seems pretty generally distributed as to localities.

(101), ∞ **P 2** occurred on about the same number of specimens as $(3\bar{1}\bar{1})$. It truncates the lateral edges of the primitive and seems to replace $(20\bar{1})$ in the modification of these parts as the two occur very seldom together: on only 6 specimens in 41. Geographically it was also very partially distributed. Only 1 specimen in 41 was from Derbyshire, while 27 were from Andreasberg. It is a remarkable fact that this face and (111) are almost never found on crystals from the former locality, while they are so largely present on those from the latter. It would be interesting to know if their crystals have passed through entirely different stages of development; or if those of the one have attained a higher development than those of the other. I find no notice of this face or $(3\bar{1}\bar{1})$ having been observed on microscopic crystals.

The remaining faces were introduced at this point to show how the series of frequency was continued. They will be noticed more specially in the table of occurrence.

I think we can be sure from the above figures that the comparative frequency of the first four faces is that above indicated, since they were not in great part from one locatily as were (111) and $(10\bar{1})$. In a perfectly unselected collection they would probably be much more frequent relatively to those just below them than the above figures show; because precisely their combinations would be those rejected.

We have found out by comparing microscopic and macroscopic crystals how the development begun on the former was continued on the latter. The predominant form of the first, (100), had in great part disappeared from the second; but the faces, which formed the first modifications of (100) on microscopic crystals are those which are met with most commonly on larger ones. This gives us the link connecting the two; and shows that in most cases the latter have had only one previons form and that the remaining forms were developed from the primitive modifications in the same way as they were developed from the primitive.

We shall now attempt to trace this development farther.

All the faces observed on microscopic crystals fall in two zones: viz, the edge-zone of the primitive and the horizontal diagonal-zone of its faces. By far the greater number of faces occurring on macroscopic crystals fall in the same zones. I counted, as above, the number of specimens on which each face occurred and added these together according to zones. The total number for all the faces was 1037, of which 961 were in the two designated zones: 544 in the rhombohedral zone, and 417 in the edge-zone. The faces of (100) were reckoned to the former (110) to the latter. There were 42 distinet forms in the two zones and 28 not in them. 906 of the faces, in 1037, belonged to the first above forms given. This figure would proportionally be still greater in an unselected collection, and also if one counted the actual number of times the several forms occurred instead of regarding each specimen as only one crystal. I do not think the case would be over stated if we said that $(2\bar{1}\bar{1})$, (110), $(20\bar{1})$, $(2\bar{1}\bar{1})$ formed 90 *percent* of the faces occurring on macroscopic crystals of calcite. However interesting its remaining forms may be, as showing the possibilities of crystallogenesis, the above figures show that they are unessential compared to the four forms named.

The axis of one of the above zones was a line already formed on the primitive crystal. That of the other was unformed but could be determined by two points actually present on it. It would be extremely difficult to see how such an imaginary line could be the efficient cause of a zone. If it be, however, that a corner where several edges meet may become a centre of modified growth, in a manner similar to that in which an edge becomes an axis of modified growth in zonal developments, it could be easily explained how this zone came to be formed. We should in fact expect such corners to be more powerful for effecting modifications than the separate edges meeting in them. This is admirably illustrated by calcite. The first observed modifications of the primitive are generally modifications not of its edges but of it corners. When a pure solution of $Ca\,CO^3$ in $H^2O + CO^2$ is taken, that is when the crystallogenic forces of calcite are allowed to act undisturbed, faces on the summit and lateral corners are almost the only ones formed. It is obvious that the law regulating such modifications is that parts modifying must preserve the symmetry of the parts modified. It follows from this that any single face modifying a corner must belong to a rhombohedron. When one such face has been formed it cuts (100) in a horizontal edge, and from this the zone may be developed in the ordinary way.

I have seen little mention of this principle of the modification of corners, but it seems to me to be of the same general character as the principle of zones and worthy to be placed beside it. It is unfortunably less useful for determing faces of crystals, but I think it will be found to be even more important in their development.

To estimate the relative importance of these two zones we must count the faces in the following manner. The faces which first determine the zone must be omitted; they were not produced by the modification of an edge. We should therefore leave out (100) from both zones and $(2\bar{1}\bar{1})$ from the rhombohedral zone, as it is probably the second determining face of the latter. (110), which is common to the two zones, should undoubtedly be counted with the edge-zone, as the ever-present striation of its surface indicates. (111) should also be omitted from the rhombohedrons as it is probably formed in the modification of the apex and not in the proper development of the rhombohedral zone. The figures thus obtained are 420 for the edge-zone and 179 for the rhombohedral zone. It would have been expected, that the edge-zone would be more important than the rhombohedral zone, if the above explanation of its development were true.

There are two other zones, of much less importance than the two above, but, which probably play the most important parts when the development of the crystals no longer takes place in the first two. They are determined by faces of $(1\bar{1}\bar{1})$ and $(2\bar{1}\bar{1})$ intersecting faces of the primitive when the two are not in the same rhombohedral zone. Since the planes of the axes, (100), are in these zones, the axes of the zones will cut two of the general axes; and the intercepts of each face of either zone on the latter will have a ratio equal to that of the intercepts of the axis of the zone. From the faces $(1\bar{1}\bar{1})$ and $(2\bar{1}\bar{1})$ we see that the ratios for the two zones are $1:\bar{1}$ and $2:\bar{1}$; the other two possible ratios, those of equality, correspond to the zone of rhombohedrons.

These two zones, which we may call the secondary zones, are the only ones in addition to the two primary zones which are determined by the intersection of the faces of $(2\bar{1}\bar{1})$, $(20\bar{1})$, (110), $(1\bar{1}\bar{1})$, (111), $(10\bar{1})$ with the faces of the primitive. It follows that several of these faces must be common to both primary and secondary zones. This renders it difficult to ascertain the relative importance of the latter; for, we are unable at present to decide in any given case whether a face has been determined by one or the other of two zones in which it falls, if either. There can be no doubt that the so-called primary zones are really those of first importance, but we may not be sure that the number obtained by counting those faces of the secondary zones not in the primary is not much too small to give a correct idea of their importance. In like manner it is difficult to determine which of the two secondary zones is the more important; for, $(21\bar{2})$, -2 R 2, by far the most frequent of the remaining faces in either, is common to both. We may, perhaps, in many cases determine to which zone it belongs by noting whether it occurs with $(2\bar{1}\bar{1})$ or $(1\bar{1}\bar{1})$. In Levy it occurs 19 times; 8 times with

$(1\bar{1}\bar{1})$ and 12 times with $(2\bar{1}\bar{1})$; 3 times with both. Each time it occurred either with one or the other. When it occurred with both, I should be inclined to count it to the zone $\left(\frac{h}{k}\,1\bar{1}\right)$ rather than the other, because the former appears, as will be seen, to be generally more important than the latter. This would give 8 times to the zone $\left(\frac{h}{k}\,1\bar{1}\right)$, and 9 times to the zone $\left(\frac{h}{k}\,1\bar{2}\right)$. Accordingly the total number of faces in the first zone, as given by Levy, would be 28 and for the second 11. This latter number would possibly have to be increased by a number of faces of $(22\bar{1})$ — R which is a rhombohedron of the zone. Perhaps the next most frequent zone in Levy is the edge zone of (110), — $^1/_2$ R. Faces occur 7 times in this. 3 of them were also faces $\left(\frac{h}{k}\,1\bar{1}\right)$. Up to the time the number of forms certainly determined in the zone $\left(\frac{h}{k}\,1\bar{1}\right)$ is 14; (100), $(1\bar{1}\bar{1})$, $(10\bar{1})$ not being included. In the zone $\left(\frac{h}{k}\,1\bar{2}\right)$ there are 7; (100), $(2\bar{1}\bar{1})$, $(20\bar{1})$ being excluded. I think there can be no doubt that the former zone has a distinct importance. With respect to the latter however we can not decide until we can determine whether certain of the faces, common to it and other zones, have not been really produced by it. A comparatively small number of such cases would be sufficient to give it a place of equal importance with zone $\left(\frac{h}{k}\,1\bar{1}\right)$.

Perhaps another zone of small but distinct importance is the edge-zone of (110), — $^1/_2$ R. The following are the known forms in it. (211), $^1/_4$ R; $(31\bar{2})$, — $^1/_2$ R 5; $(52\bar{3})$, — $^1/_2$ R 4; $(41\bar{3})$, — $^1/_2$ R 7; $(51\bar{4})$, — $^1/_2$ R 9; $(11\ 2\bar{9})$, — $^1/_2$ R 10; $(71\bar{6})$, $^1/_2$ R 13. The number of faces is considerable, but it is necessary to determine whether they have generally been formed by the modification of the edge of — $^1/_2$ R, in order to establish the importance of the zone.

The most interesting facts I have been able to collect concerning the combinations are the following. All the figures are from Levy.

The number of combinations of (110), $(20\bar{1})$, $(2\bar{1}\bar{1})$, $(1\bar{1}\bar{1})$, (111) was 153 in 346. It is however precisely these combinations which would be rejected in selecting a collection. I think the true proportion of them would be $^2/_3$ if not higher. This shows that in the majority of cases the first modifications of (100) proceed so far as to cause its disappearance but that no further development takes place. If we add (100), (101) to the above the faces the number of combinations becomes 168 in 346.

The number of combinations of faces belonging to the primary zones was 268 in 346 or about $^3/_4$. The true proportion would of course be higher.

The number of combinations of faces in the primary and secondary zones was 306 in 346 or about $^7/_8$.

The following figures will give are idea of the comparative frequency of the principal combinations. The total number of specimens was 346.

(110); $-\,^1/_2\,R$	19	
$(1\bar{1}\bar{1})$; $-\,2\,R$	15	
(111), $(2\bar{1}\bar{1})$; $0\,R$, $\infty\,R$	15—12	from Andreasberg.
$(25\bar{1})$; $R\,3$	14—8	„ Derbyshire.
$(2\bar{1}\bar{1})$, (110), $(20\bar{1})$; ∞R, $-\,^1/_2 R$, $R\,3$	9—6	„ „
(110), (111), $(2\bar{1}\bar{1})$, $(20\bar{1})$; $-\,^1/_2\,R$, $-\,2\,R$, $\infty\,R$, $R\,3$ . . .	8— all	„ „

I think that $(110)\,(2\bar{1}\bar{1})$ will be found to be the most frequent of all combinations. From Pribram, Bohemia, where it occurs in such great numbers, there was only one specimen.

c) Twins.

Law I. Twin-plane (110); — $^1/_2$ R. The two members have two cleavage planes common: also the axes of 1 edge-zone of primitive and 1 rhombohedral zone. The paper of G. Rose's on the *"hohlen Canäle"* treats of the occurrence of lamellae disposed according to this law in the interior of crystals. Occurring in this way they are the most frequent of all kinds of twins, though it is somewhat rare that both members are developed as separate crystals. It is obvious that $(10\bar{1})$ might also be called the twin-plane for this law. I know of no mention of twins with it as the plane of attachment.

Law II. Twin-plane (100); R. The two numbers have one cleavage plane and edges and diagonals of corresponding face of primitive common. The members are attached by 100.

Law III. Twin-plane (111); 0 R. The rhombohedral zones of the two members have common axes. $(2\bar{1}\bar{1})$; ∞ R may also be called the twin-plane. The members are attached both by (111) and $(2\bar{1}\bar{1})$. These twins are the most frequent with both members developed.

Law IV. Twin-plane $(1\bar{1}\bar{1})$; — 2 R. The members have two clinodiagonals of the primitive common. These are much the rarest of all twins. All the notices of them I have seen are those of Scheerer, von Kobell (twice), vom Rath (n. F. 8.) and Scharff. Three of the combinations were $(20\bar{1})$, attached so that x-edges on the two members were parallel: the other was a cleavage block.

Sella in his „Studi etc.“ has given two additional kinds of twins with planes $(22\bar{1})$; — R and (441); $^1/_2$ R. These have been critically discussed by vom Rath (no. 20), who thinks the twin planes should properly be (100); + R and $(1\bar{1}\bar{1})$ — 2 R. The corresponding laws in each case could easily be confounded, for (100) almost symmetrically truncates the edge 111: $2\bar{1}\bar{1}$, and consequently the faces 100 and $1\bar{2}\bar{2}$ are almost perpendicular to one another. So likewise are 441 and $1\bar{1}\bar{1}$.

The reappearance of (110), (100), (111), ($2\bar{1}\bar{1}$), ($1\bar{1}\bar{1}$) as twin-planes is another proof the extreme importance of these forms.

A point which it would be interesting to decide but upon which the existing observations are most meagre is, whether the members of twins are united to each other with the same degree of accuracy as the different parts of the same simple crystal are united to one another. Only measurements would be necessary for this purpose but I think there are scarely a dozen recorded measurements of twins in all the works above given.

III. The Tables.

There are two parallel tables, one of occurence and one of angles. In these are contained all the forms which are determined with a reasonable degree of certainty. A list of uncertain and incorrect forms is appended to these. In each of the tables the forms of the primary and secondary zones have been placed together and after these the remaining forms. To complete the list of forms in each zone it has been necessary to repeat those forms common to two zones. The principle of classification according to which, the forms not in the given zones are arranged, is, in Naumann's nomenclature, the height of the form. It is perfectly arbitrary but it is intended simply for the purpose of enabling one easily to find the forms. Expressed in Miller's symbols, the classification is according to values of the function.

$$\frac{h - 1}{h + k + 1}.$$

According to this latter the pyramids also find proper places.

In the table of occurrence, where not otherwise indicated, I have given, with approximate completeness, references to all recorded observations of the several faces known to me. The following abbreviations have been used. Z = Zippe, gr = Gruppe, H = Hessenberg, the appended Roman number is the *volume* of the transactions in which the required paper is to be found, R = vom Rath, the appended Arabic number refers to the required *Mittheilung*. The remaining abbreviations and references used in the table and other parts of the work will be easily understood by referring to the list of literature.

The table of angles has been calculated from the fundamental angle 100 V 010 = 105° 5′. This is the value to be found in most of the books. The following are the results of the principal measurements. Wollaston, Malus and Biot's determinations gave 105° 5′. Kupfer obtained 105° 4·5′. Vom Rath (19 c) obtained 105° 4·8′. Kokscharow 105° 4′. Breithaupt, who measured this angle on almost 3000 specimens, says that its value, for perhaps $^{2}/_{3}$ of the cases, was between 105° 8′ and 105° $8^{3}/_{4}$′. This angle deviates too widely from the above to be attributed to accidental errors, but I think it probably due to a personal equation of error. The measurements were made with a Wollaston's goniometer. It seems to me that no instrument could be more favorable for the development of such.

The variation of this angle with the temperature is, according to Mitscherlich and Breithaupt, about 5″ for 1 degree centigrade.

Instead of the apical angles of the rhombohedrons and the angles of the three edges of skalenohedrons I have given the halves of these angles, as they are always used in calculations. The remaining angles given are useful in calculations.

Rhombohedrons.

Table of occurrence.

1.	100 R	See p. 19.
2.	711 $^{2}/_{3}$ R	In Levy's fig. 134 incorrectly given as a^3 (311), It is in the zone 401 : 111 and from this it may be determined.
3.	511 $^{4}/_{7}$ R	Zippe gives this form as doubtful. It is satisfactorely determined in however in a comb. from the Maderaner Thal (H. III. 267.)[1]).
4.	411 $^{1}/_{2}$ R	Only known in Levy's var. 153. In the fig. it is lettered p. but is drawn as (411). The text gives (411).
5.	311 $^{2}/_{5}$ R	In a comb. from the Maderaner Thal (H. III. 267.)[1]).
6.	211 $^{1}/_{4}$ R	According te Zippe it occurs as a simple form and in comb. He gives but one comb. however and that is from the inaccurate Bournon. I find it mentioned by no other author.
7.	111 0 R	See p. 21.
8.	11118 — $^{1}/_{10}$ R	In a comb. from Maderaner Thal. (H. III. 267.)[1]).
9.	221 — $^{1}/_{5}$ R	Given twice by Levy: var 2, simple form from Offenbanya, Var. 116, comb. from Andreasberg. According to Frenzel (Min. Lex. p. 50) it occurs in the socalled „Papierdrusen" at Schneeberg in Saxony and also at Gersdorf.
10.	992 — $^{7}/_{20}$ R	Peters (N. Jahrb. 1861) in a comb. from Czalathna. He gave the measured angle 299 V 211=109° 10' but did not calculate the result.
11.	771 — $^{2}/_{5}$ R	In a comb. from Freiberg determined by Dauber (R. 19 h.)

1) Hessenberg, in a subsequent correction, gave this specimen as dolomite from Binnenthal.

12.	110 $-\frac{1}{2}R$	See p. 20.
13.	$88\bar{1}$ $-\frac{8}{5}R$	According to Zippe in a comb. with $(2\bar{1}\bar{1})$.
14.	$55\bar{1}$ $-\frac{2}{3}R$	According to Zippe in a comb. of Hausmann's: $(55\bar{1})$ $(2\bar{1}\bar{1})$ $(\bar{1}\bar{1}\bar{1})$. In several comb. from Agaëte. (H. VII. 265.)
15.	$33\bar{1}$ $-\frac{4}{5}R$	Levy fig, 10. — Breithaupt fig. 213. — H. III. 267 on dolomite. — Sella, Studi. fig. 71. — Peters, N. Jahrbuch, 1861. — Z. fig. 65. Hauy gevis it in 11 combb.
16.	$55\bar{2}$ $-\frac{7}{8}R$	Levy, figg. 113 a. 121. — R. 19 g: from Maxen near Dresden.
17.	$22\bar{1}$ $-R$	One of the less seldom forms. Levy gives 10 comb. a. Zippe also a number.
18.	$55\bar{3}$ $-\frac{8}{7}R$	Z. fig. 90. — H. III. 265. 2 comb. — H. III. 266. — R. 19 d. — R. 83. — Should replace — $\frac{9}{8}$ R in H. VII. figg. 3 u. 6. See app. I.
19.	$1111\bar{7}$ $-\frac{6}{5}R$	Naumann, Pogg. Ann. Bd. XVI. Andreasberg. — H. IV. 6. Bleiberg.
20.	$33\bar{2}$ $-\frac{5}{4}R$	Levy, var. 162. — Z. fig. 11, gr. 7 a. — H. VIII. (uncertain) — R. 19 c. — H. IV. 13, in a comb. of Volgers. — Hauy figg. 47. 51. 53. 76. 137.
21.	$77\bar{5}$ $-\frac{4}{3}R$	Hessenberg (VII. 15) on Crystal from Agaëte.
22.	$43\bar{3}$ $-\frac{7}{5}R$	This form was first observed by Weiss. (Abhand. Berliner Akad. 1836.) It lies in the Zone determined by the alternate faces of $(20\bar{1})$. Levy, varr. 100. 115. 185. 176. — R. 5 n. F. from Bergen Hill. — Websky (Tschermak, Min. Mitth. 1872. p. 63) from Striegau.
23.	$55\bar{4}$ $-\frac{3}{2}R$	A some what frequent form. Levy mentions 20 Specimens with it, Occurs simple as will as in comb. Hauy gives it in 7 combb.
24.	$66\bar{5}$ $-\frac{11}{7}R$	Z. fig. 67, from Alston Moore. — H. X. 17. Andreasberg.
25.	$77\bar{6}$ $-\frac{13}{8}R$	Zippe gives 11 combb. all from Bournon. — H. VII. Agaëte. — H. IV. 190. Andreasberg.

26.	$11\bar{1}$ $-2R$	Sec. p. 20.
27.	$13\,13\,\bar{14}$ $-\frac{9}{4}R$	Zippe gives this form as determined from measurements of Bournon and says it is uncertain. Zepharovich gives it (Min. Lex. I. p. 86) from Böhmisch-Leipa as a simple form.
28.	$77\bar{8}$ $-\frac{5}{2}R$	Zippe, fig. 90: from Andreasberg. — R. 20: from Elba a. from Iberg.
29.	$55\bar{6}$ $-\frac{11}{4}R$	Z. gr. 13, 4 combb. of Bournon's from Derbyshire. Levy varr. 38. 97. 173.
30.	$33\bar{4}$ $-\frac{7}{2}R$	Z. gr. 14, 5 combb. of Bournon's. 1 of Haidinger. a. 1 other. Levy. Varr. 36. 127. 156. 173. 189. — R. 19 a. from Lake Superior. — Zepharovich. (Min. Lex. II. 76) from Lölling in Kärnthen. Hauy, figg. 49. 144.
31.	$55\bar{7}$ $-4R$	Z. fig. 36 from Derbyshire. — H. IV. with + 4 R from Bleiberg. VIII. p. 37 from Hauy, figg. 147. 64.
32.	$11\,11\,\bar{16}$ $-\frac{9}{2}R$	R. 19, d: from Andreasberg. Truncates X-Edge of R $^{19}/_{3}$.
33.	$22\bar{3}$ $-5R$	A not infrequent form. Levy gives 10 combb. a. Z. 8 additional ones. — Kokscharow fig. 16. — Zepharovich: 1) I. p. 16: from Nieder Einsiedel in Silesia. 2) II. p. 76: from Lölling in Kärnthen. Hauy gives it in 9 combb.
34.	$77\,\bar{11}$ $-6R$	H. X. 13 from Island.
35.	$33\bar{5}$ $-8R$	Levy var. 136. — Z. gr. 10 h, in 4 combb. from Maria Trost near Gratz a, in fig. 14 from England. — Hauy, fig. 131, a. comb. 150.
36.	$44\bar{7}$ $-11R$	Z. (gr. 10 a.) erroneously gives this form as in Levys fig. 49. — Sella, Studi. fig. 71. — Kokscharow fig. 21. — H. VII. from Agaëte.
37.	$55\bar{9}$ $-14R$	Z. gr. 19, gives 23 comb. 14 of which are from Andreasberg. Levy gives 11 specimens, 6 of which are from Andreasberg. — Z. in the remarks to his fig. 44, in which he corrects Levys fig. 86. His correction is based on apparent zone in the figure. The Edges of the zone are too short, however, for the parallelism to be cer-

		tainly. – R. 20, f. — Kokscharow fig. 27. — Sella, Studi. figg. 34 a. 71. Hauy, figg. 34. 71. 90. 91. 95. 109. 149. This one of the less seldom forms.
38.	$\frac{11\overline{2}}{\infty\ R}$	See p. 19.
39.	$\frac{99\,\overline{19}}{28\ R}$	Determined by Zippe, gr. 21 a. fig. 49, on a specimen from Alston Moor.
40.	$\frac{17\,17\,\overline{37}}{18\ R}$	The indices of this form are quite large, but it is accurately determined. (R. 19, a.)
41.	$\frac{55\,\overline{11}}{16\ R}$	This form was unknown to Hauy a. Levy. Zippe gives it in 5 combb. 3 from Derbyshire a. 2 from Dufton. — Scharff (N. Jahrbuch 1861 p. 542 u. 544) gives it in 2 comb. one from Maderaner Thal a. the other from Andreasberg. Frenzel (Min. Lex. p. 52) mentions it on crystals from Niederrabenstein near Chemnitz and from Oschatz in Saxony.
42.	$\frac{44\overline{9}}{13\ R}$	Z. gr. 18 a, gr. 5 c gives 5 comb. One of them, 13 R, $^1/_2$ R, as occurring at 5 different localities. Levy calatogues 16 specimens with it. — R 19 c from Hausach in Baden. — Descloizeaux, Manuel, p. 267. Andreasberg. Hauy, figg. 34. 126. 127. 129. 154. 155.
43.	$\frac{33\overline{7}}{10\ R}$	This form was first observed by Descloizeaux (Atlas fig. 268) on crystals from Iceland. Hessenberg also found it on crystals from the same source (VI. VIII. X.). He first thought it be 9 R. — R 76: from Lake Superior.
44.	$\frac{22\overline{5}}{7\ R}$	Z. fig. 42. — Levy, varr. 163. 174. 179. — Sella, Studi. fig. 71: from Traversella. — Hauy, comb. 150.
45.	$\frac{55\,\overline{13}}{6\ R}$	Sella, Studi. fig. 32: from Traversella.
46.	$\frac{33\overline{8}}{^{11}/_2\ R}$	Levy, fig. 49. from Iberg. It is erroneously given in text as (335). H. IV. p. 13 gives a comb. copied from Volger. Zippe omits it.
47.	$\frac{44\,\overline{11}}{5\ R}$	R. 19 d. from Andreasberg. Truncates Y-Edge of R $^{19}/_3$.
48.	$\frac{11\overline{3}}{4\ R}$	See p. 21.

49.	227 3 R	This only accurate observation of this form is by Sella, Studi. figg. 24. 25. 26: from Traversella.
50.	$11\bar{4}$ $\overline{{}^5/_2}\,R$	Z. gr. 12: from the Harz. — Levy, varr. 96 a. 99. — R. 19, a. from Lake Superior. — Kokscharow, fig. 10. — Zepharovich (Min. Lex. p, 76), Sulzbach, Graben, Steirmark. — Peters (N. Jahrb., 1861. p. 439) from unknown locality.

Skalenohedrons.

Table of occurence.

I. Edge-zone of primitive.

51.	320 $-{}^1/_5\,R\,3$	Hauy, fig. 38. Descloizeaux fig. 267 Andreasberg.
52.	530 $-{}^1/_8\,R\,5$	Levy, fig. 26 Strontian, Scottland. — Scharff. (N. Jahrbuch 1862, p. 699), from Lake Superior.
53.	210 ${}^2/_3\,P\,2$	Hauy figg. 127. 146 a. comb. 134. — Levy, varr. 63. 89. 104. 151. — H. IV. Maderaner Thal. — Sella, Studi. fig. 41. Andreasberg.
54.	730 ${}^1/_{10}\,R\,7$	Z. gr. 12, 28 a, 28 e, 57: 2 comb. after Hausmann and 3 after Bournon. — Z. fig. 79. — H. VII. Agaëte. — R. 24: from the Nahe. — Peters, N. Jahrbuch, 1861. p. 437. Andreasberg.
55.	11 40 ${}^1/_5\,R\,{}^{11}/_3$	R. 19 a. Lake Superior. — H. X. Andreasberg.
56.	310 ${}^1/_4\,R\,3$	The most frequent form which bevels the apical Edges of the primitive. Hauy gives it in 9 combb. Levy gives it on 30 Specimens, 24 of which were from Derbyshire. On all the Derbyshire Specimens and some of the others it occurred with $(20\bar{1})$ R 3.
57.	720 ${}^1/_8\,R\,{}^7/_3$	Sella, Quadro. 117: from Andreasberg.
58.	410 ${}^2/_5\,R\,2$	Hauy, figg. 26. 112. 154. 155. Levy found this form, only on 1 spec. var. 7. — R. 19 e, a. 83. — Z. gr. 28, h,

		46 a, figg. 5. 20. 31. 80. — Descloizeaux f. 266. — H. IV, a VIII. Of those faces of the zone above (110) this I think is the most frequent after (310).
59.	920 ${}^{5}/_{11}\,R\,{}^{9}/_{5}$	R. 5 n. f. from Bergen Hill.
60.	510 ${}^{1}/_{2}\,R\,{}^{5}/_{3}$	Z. figg. 7, a. 81: from St. Pancraz near Prag, and Andreasberg. Scharff, N. Jahrbuch 1862. p. 70.
61.	610 ${}^{4}/_{7}\,R\,{}^{3}/_{2}$	Levy, varr. 121. 146. 190. Zippe, fig. 61, a. gr. 27 e. Hauy, fig. 37.
62.	710 ${}^{5}/_{8}\,R\,{}^{7}/_{5}$	Z. fig. 33 after Hausmann. Andreasberg. — H. IV. Andreasberg.
63.	810 ${}^{2}/_{3}\,R\,{}^{4}/_{3}$	R. 76. Lake Superior. — H. X. Andreasberg.
64.	910 ${}^{7}/_{10}\,R\,{}^{9}/_{7}$	Z. fig. 50. Andreasberg. — R. 19, a. Lake Superior.
65.	$14\,1\,0$ ${}^{4}/_{5}\,R\,{}^{7}/_{6}$	R. 24 from the Nahe. The measurement of this form was only approximative on account of its being corroded. It is therefore not certainly distinguished from the forms (13,10) a. (15,10), they being only about 20 of angle of angle distant on either side from (14,10).
66.	$17\,0\,\bar{2}$ $R\,{}^{19}/_{15}$	Z. fig. 24. Bleiberg. — H. VIII. Bleiberg. — Schnorr, Zwickau. — R. 83. Ahrenthal, Tyrol.
67.	$70\bar{1}$ $R\,{}^{4}/_{3}$	Z. fig. 51. This determination is uncertain. H. IV. Matlock, a. Andreasberg.
68.	$60\bar{1}$ $R\,{}^{7}/_{5}$	Hauy, fig. 108. — Levy, fig. 54. Andreasberg. — Z. fig. 84.
69.	$50\bar{1}$ $R\,{}^{3}/_{2}$	Hauy, fig. 36. 110. 154. — Levy, fig. 65. Z. figg. 8. 55, a. gr. 61, a. — Scharff, N. Jahrbuch. 1862, p. 710. — Kokscharow, fig. 10.
70.	$40\bar{1}$ $R\,{}^{5}/_{3}$	Hauy, figg. 104. 125. 126. 127. 135. 143. — Levy, varr. 71. 117. 118. 151. 172. — Zippe, figg. 3. 9. 81. — Scharff, N. Jahrbuch 1862. pp. 703. 710.
71.	$70\bar{2}$ $R\,{}^{9}/_{5}$	Levy, figg. 39. 40. — H. IV. Andreasberg. — Descloizeaux fig. 267, from Andreasberg a. p. 106 from Pribram. — Scharff, N. Jahrbuch 1862. p. 699.

72.	$30\bar{1}$ R 2	Monteiro, Jour. d. Mines 1813 II. p. 194. — Hauy, figg. 28. 86. — Z. gr. 61 a, figg. 85. 86: 6 combb. all from Braunsdorf near Freiberg. — H. X. Agaëte a. Iceland.
73.	$50\bar{2}$ R $^{7}/_{3}$	Hauy, fig. 63. comb. 134. — Z. gr. 10 f.
74.	$20\bar{1}$ R 3	See p. 20.
75.	$170\bar{9}$ R $^{13}/_{4}$	Z. fig. 65.
76.	$70\bar{4}$ R $^{11}/_{3}$	Hauy, figg. 15. 63. 141. — Z's. gr. 30 is taken entirely from Bournon. — Zepharovich, Sitzungsberichte der Wiener Akademie. Bd. 54.
77.	$50\bar{3}$ R 4	Levy, var. 34. — Z. gives 1 comb. with $^{5}/_{3}$ R. $^{7}/_{5}$ taken from Hausmann. Sella, Studi, figg. 3. 22. Traversella. — R. 49. Alston Moor, 83 Ahrenthal, Tyrol.
78.	$80\bar{5}$ R $^{13}/_{3}$	Z. fig. 91 gr. 66 a. — R 83 Ahrenthal. — H. VI. Iceland. — Kokscharow, figg. 12. 16.
79.	$30\bar{2}$ R 5	Of the faces of the Zone below (100) this is the most frequent after $(20\bar{1})$. I judge it to be of about the same degree of frequency as (310). Hauy gives 14 combb. Levy 15.
80.	$190\overline{13}$ R $^{16}/_{3}$	H. IV. from Andreasberg. This, the only determination of this form, is not perfectly satisfactory.
81.	$100\bar{7}$ R $^{17}/_{3}$	Hauy, comb. 100. — H. III. from Mattock. — R. 19, c: from Arendal a. 5: n. F. from Bergen Hill.
82.	$110\bar{8}$ R $^{19}/_{3}$	R. I9 d Andreasberg.
83.	$40\bar{3}$ R 7	Monteiro Ann. d. Mines, V. 3. 1820. — Levy gives 18 combb. 13 of which are from Andreasberg. Hauy does not give it. — Z. grr. 1 a, 15 c a. f, 42, 46, figg. 60, 63. This form ranks in frequency after $(30\bar{2})$ among the faces below (100).
84.	$50\bar{4}$ R 9	Hauy gives this form in 8 combb. Levy, in 4. — Z. grr. 15 d, 34 a, 46, fig. 45, 46. — H. III. — R. 16 a, 20 a.
85.	$60\bar{5}$ R 11	A form unknown to Hauy. Levy gives it in 7 combb. Zippe adds only his fig. 65 to these. — R. 20 d.

86.	$13\,0\,\overline{11}$ R 12	Z. gives in gr. 36. 9 combb. of this form taken from Naumann, a. Hausmann. It has not, however, been otherwise observed, and I am doubtful of the correctness of the determinations. Z. does not say where the original observations are to be found and I was, therefore, unable to compare them with the calculated angles. If this form exist its faces should be quite perfect in order to be distinguishable from R. 11, a. R. 13 which are only about 30′ of angle distant from it on either side.
87.	$70\bar{6}$ R 13	Levy, figg. 7. 20: from the Faro Islands. Z. figg. 66. 68: Alston Moor. — R. 16 c: Arendal. — H. III. Matlock.
88.	$90\bar{8}$ R 17	Websky (Tschermak, Min. Mit. 1872) from Striegau in Silesia.
89.	$10\bar{1}$ ∞ P 2	See p. 22.

II. Zone $\left(\frac{h}{k}\ 1\ \bar{1}\right)$.

90.	$51\bar{1}$ $^2/_5$ R 3	Levy, fig. 149 from Derbyshire. The symbol of this face erroneously printed e^8 in the figure, and e_3 in the text. Weiss determined it however by two Zones.
91.	$41\bar{1}$ $^1/_4$ R 5	Levy, fig. 113. from Dept. de l'Isere. France.
92.	$31\bar{1}$ $^4/_3$ P 2	Levy, varr. 74 a. 144. from Andreasberg a. Derbyshire. H. IV. from Bleiberg a. VIII. from Iceland.
93.	$52\bar{2}$ $-\,^1/_5$ R 7	Z. gr. 63 a. in comb. with $(20\bar{1})$ $(2\bar{1}\bar{1})$ $(10\bar{1})$ after Hausmann.
94.	$73\bar{3}$ $-\,^2/_7$ R 5	Z. fig. 90 and in comb. with $(10\bar{1})$ (Hausmann).
95.	$21\bar{1}$ $-\,^1/_2$ R 3	Levy, varr. 149. 150. 165. 190. — Z. gr. 55 a a. fig. 89. — Sella, Studi. fig. 34.
96.	$95\bar{5}$ $-\,^2/_3$ R $^7/_3$	Naumann, Pogg. Ann. XIV. p. 236.
97.	$32\bar{2}$ $-$ R $^5/_3$	Hauy, fig. 138. Levy, varr. 161. 162. R 19 d a. 20 a. Andreasberg.
98.	$75\bar{5}$ $-\,^8/_7$ R $^3/_2$	Wimmer. Bericht über die dritte General-Versammlung d. Clausthaler naturwissenschaftlich. Vereins, Maja 1854.

99.	$43\bar{3}$ / $-\,^{5}/_{4}R\,^{7}/_{5}$	Levy, varr. 44. 75. 135. 165.
100.	$23\bar{3}$ / $-2R\,^{3}/_{2}$	Levy, varr. 82. 87.
101.	$35\bar{5}$ / $-2R\,^{5}/_{3}$	Z. gr. 42. 66. 11 combb. from Haidinger: all from Derbyshire; gr. 10 a. from Island; gr. 46 b. from the Harz. This is the second most frequent form of the zone.
102.	$21\bar{2}$ / $-2R2$	The most frequent skalenohedron of the zone. It is characteristic of the calcite from Alston Moor. Levy gives it on 20 Specimens, 11 of which are from this locality. The total number of specimens from the locality was 31, consequently (212) occured on about one third of them. Hauy gives it in 7 combb. Zippe in 7. Also, Kokscharow fig. 21. — H. IV. — Descloizeaux fig. 270. — R 19 a, 196. 48.
103.	$31\bar{3}$ / $-2R3$	Z. figg. 23. 44. 57. 73. 88. 89. gr. 28 e, 44. Zippe for reasons explained in the remarks to his fig. 32 has substituted this form for ($d^{1}/_{2}$ $d^{1}/_{4}$ $b^{1}/_{3}$) of Levy's fig. 143 a. for $1^{2}/_{3}$ of fig. 155 of the same author. — H. IV. Bleiberg. This determination is doubtful.
104.	$41\bar{4}$ / $-2R4$	Z. fig. 58.

III. Zone $\left(\frac{h}{k}\ 1\ \bar{2}\right)$.

105.	$41\bar{2}$ / $4R\,^{8}/_{3}$	Z. fig. 6. — Levy, fig. 7 Pl. XII. on dolomite. — H. VII. Agaete. According to Scharff, Breithaupt determined this form on the nuclii of certain crystals from Tharand in Saxony.
106.	$31\bar{2}$ / $-\,^{1}/_{2}R5$	Z. gr. 10, figg. 35. 59. 75. 76. — H. IV. Matlock.
107.	$52\bar{4}$ / $-R3$	Hauy, fig. 96. — Levy, fig. 25. — Z. gr. 63.
108.	$32\bar{4}$ / $-5R\,^{7}/_{5}$	Levy, varr. 152. — Sella, Studi. fig. 34.
109.	$43\bar{6}$ / $-8R\,^{5}/_{4}$	Z. fig. 66. R. 83. Ahrenthal.

IV. Forms not in the foregoing Zones.

110.	$11\,4\,\bar{3}$ $^{7}/_{6}\,P\,2$	Descloizeaux substitutes this form for $^{10}/_{9}$ P 2 (H. IV. Maderaner Thal). $11 4 \bar{3}$ V $10\bar{1}$=134° 55′ 18″: measured 135°.
111.	$10\,7\,\bar{6}$ $-{}^{10}/_{11}\,R\,^{8}/_{5}$	See Appendix I: from Lake Superior.
112.	$16\,10\,\overline{11}$ $-\,R\,^{9}/_{5}$	Sella, Studi. fig 34. Traversella.
113.	$15\,7\,\bar{9}$ $-{}^{8}/_{13}\,R\,3$	See Appendix I: from Lake Superior.
114.	$32\,14\,\overline{19}$ $-{}^{5}/_{19}\,R\,^{17}/_{5}$	See Appendix I: from Lake Superior.
115.	$12\,2\,\bar{5}$ $^{1}/_{3}\,R\,^{17}/_{3}$	See Appendix I: from Agaëte.
116.	$25\,11\,\overline{15}$ $-{}^{4}/_{7}\,R\,^{10}/_{3}$	See Appendix I: from Lake Superior.
117.	$50\,21\,\overline{30}$ $-{}^{22}/_{41}\,R\,^{40}/_{11}$	See Appendix I: from Lake Superior.
118.	$5\,2\,\bar{3}$ $-{}^{1}/_{2}\,R\,4$	Hauy, fig. 26. — Descloizeaux, fig. 270. — R. 19 a. 76. 83. 90.
119.	$6\,3\,\bar{4}$ $-{}^{4}/_{5}\,R\,^{5}/_{2}$	Levy fig. 75. Weiss, and after him Zippe regarded this as a doubtful form, because no zones are present by which the determination could be checked.
120.	$7\,1\,\bar{3}$ $^{2}/_{5}\,R\,5$	See Appendix I: from Agaëte.
121.	$9\,1\,\bar{4}$ $^{1}/_{2}\,R\,^{13}/_{3}$	See Appendix I: from Agaëte.
122.	$6\,4\,\bar{5}$ $-{}^{7}/_{5}\,R\,^{11}/_{7}$	Sella, Studi. fig. 34. Traversella.
123.	$5\,3\,\bar{4}$ $-{}^{5}/_{4}\,R\,^{9}/_{5}$	Z. gr. 10 f., 47, a. figg. 53. 69. 82.

124.	$95\bar{7}$ $-{}^{8}/_{7}\,R\,2$	Levy, varr. 126. 192. 193. 197. — Z. fig. 70.
125.	$42\bar{3}$ $-R\,{}^{7}/_{3}$	Naumann, Pogg. Ann. XIV. — Z. fig. 28. On the determination of this form in Levy's fig. 143 see the remarks to Zippe's fig. 32.
126.	$73\bar{5}$ $-{}^{4}/_{5}\,R\,3$	Hauy, figg. 101. 130. Levy gives it in 12 combb. — Z. gr. 28 g, 62 a, 62 f, figg. 88. 89. — Descloizeaux, fig. 270. — R. 19 a, 19 c. It is perhaps the most frequent form not in the principal zones. In 10 of Levy's 12 combb. it occurs as a bevelling on the X-Edge of $20\bar{1}$.
127.	$10\,5\,\bar{8}$ $-{}^{8}/_{7}\,R\,{}^{9}/_{4}$	Levy, fig. 138. Naumann's Symbol for this form is incorrectly given in Zippe as ${}^{5}/_{4}\,R\,{}^{9}/_{4}$. Descloizeaux somewhat arbitrarily, substitutes for it ($95\bar{8}$), because the latter falls in certain zones on the general chart of the calcite faces.
128.	$51\bar{3}$ ${}^{8}/_{3}\,P\,2$	Z. gr. 10 f, one comb. from Mohs.
129.	$13\,1\,\bar{7}$ $-{}^{4}/_{7}\,R\,5$	H. IV. Derbyshire.
130.	$35\,17\,\overline{32}$ $-{}^{31}/_{21}\,R\,{}^{67}/_{31}$	R. 90: from Elba. This form is far from simple, but it seems impossible to substitute any other for it.
131.	$23\,11\,\overline{21}$ $-{}^{20}/_{13}\,R\,{}^{11}/_{5}$	R. n. f. 5: from Bergen Hill.
132.	$41\bar{3}$ $-{}^{1}/_{2}\,R\,7$	Kokscharow, figg. 12. 16.
133.	$17\,10\,\overline{18}$ $-{}^{7}/_{3}\,R\,{}^{5}/_{3}$	H. VIII. Iceland.
134.	$71\bar{5}$ $4\,P\,2$	Hauy, fig. 122. — Levy, varr. 137. 178. 193.
135.	$72\bar{6}$ $-R\,{}^{13}/_{3}$	Levy, fig. 115.
136.	$51\bar{4}$ $-{}^{1}/_{2}\,R\,9$	Levy, var. 193.
137.	$21\,3\,\overline{11}$ ${}^{16}/_{7}\,R\,2$	Levy, figg. 23. 24. — H. III. Derbyshire.

138.	$81\bar{6}$ $^{14}/_{3}$ P 2	Z. fig. 51.
139.	$15\,\overline{19}$ $^{8}/_{5}$ R 3	Hauy, figg. 133. 152. — Levy, varr. 128. 180 181. — Z. 15 h, figg. 34. 60. Of the forms not in the principal zones this is probably the most frequent after — $^{4}/_{5}$ R 3.
140.	$82\bar{7}$ — R 5	Hauy, fig. 147.
141.	$112\bar{9}$ — $^{1}/_{2}$ R 10	H. IV. from Derbyshire.
142.	$13\overline{27}$ $^{5}/_{2}$ R 2	Zippe, fig. 5, gr. 47 a. Descloizeaux proposes to substitute for this form $(31,5,\overline{17})$ but for reasons which seem insufficient.
143.	$113\bar{5}$ 4 R $^{4}/_{3}$	H. VII. Agaëte. — R. 19 a. Lake Superior.
144.	$91\bar{7}$ $^{16}/_{3}$ P 2	R. 19 d from Andreasberg. — H. VII from Agaëte.
145.	$12\,3\,\bar{1}\bar{1}$ — $^{5}/_{4}$ R $^{23}/_{6}$	R. 24: from the Nahe.
146.	$11\,\overline{17}$ 2 R 3	Zippe, fig. 57. Derbyshire.
147.	$10\,1\bar{8}$ 6 P 2	Hauy, figg. 4. 72. 144. 152. — Levy, varr. 79. 86. 181. 198. — Z. fig. 50.
148.	$71\bar{6}$ — $^{1}/_{2}$ R 13	Levy, varr. 84. 120.
149.	$95\,\overline{11}$ — 4 R $^{5}/_{3}$	H. VI. VII. VIII: all three from Iceland. — R. 19 a. Lake Superior, 24 from the Nahe, 76 Lake Superior. — Descloizeaux, fig. 268. Iceland.
150.	$35\,\bar{7}\,\overline{20}$ $^{29}/_{8}$ R $^{55}/_{29}$	I have substituted this form for Zepharovich's quite impossible $^{19}/_{5}$ R $^{19}/_{10}$ $(1373\ \overline{280}\ \overline{97}3)$. (Wiener Sitzungsberichte Bd. 54.) Calculated for $(35,\overline{7,20})$ Mean of measurements. X = 87° 32′ 26″ . . . . 87° 43′ Y = 154° 14′ 34″ . . . . 154° 14′ This form occured with (704) R. $^{11}/_{3}$. The symbol of the substitute may be written $7\ \overline{^{7}/_{5}}\ \bar{4}$. It will at once be evident that, since $^{1}/_{7}$ & — $^{1}/_{4}$ are common intercepts, these

		two and a face of the primitive lie in a zone; or in other words, $(70\bar{4})$ and $(7\ \overline{^5/_7}\ \bar{4})$ have a common mode of arrangement in one plane of cleavage.
151.	$\frac{5\bar{1}\bar{3}}{4R2}$	Levy, varr. 76. 125. 163. 164. 179. — Zippe, figg. 10. 58. — H. III. Rossie; VII. Agaëte.
152.	$\frac{13\,1\,\overline{11}}{8\bar{P}2}$	R. 19 e. from Hausach in Baden a. 19 b. from Alston Moor.
153.	$\frac{7\bar{1}\bar{5}}{4R3}$	Z. fig. 29.
154.	$\frac{7\bar{2}\bar{5}}{\infty R2}$	Hauy, fig. 139.
155.	$\frac{5\bar{1}\bar{4}}{\infty R3}$	Z. grr. 3 a, 56. This is the only form, which I have admitted on the sole authority of Bournon. It is difficult to understand how one could be mistaken in is determination.

156.	$\frac{65\,30\,\overline{39}}{-^{17}/_{28}R^{52}/_{17}}$	See Appendix I.

Rhombohedrons.

Table of Angles.

No.	Symbol of Miller.	Symbol of Naumann.	1/2 (Apical Angle).	Inclination of face to Axis.	Inclination of Edge to Axis.
1	100	R.	52° 32′ 30″	45° 23′ 26″	63° 44′ 46″
2	711	2/3 R	61 35 6	56 40 10	71 47 53
3	511	4/7 R	64 50 6	60 35 25	74 15 35
4	411	1/2 R	67 28 30	63 44 46	76 8 47
5	311	2/5 R	71 27 57	68 28 0	78 50 22
6	211	1/4 R	78 1 56	76 8 47	82 58 14
7	111	0 R	90 0 0	90 0 0	90 0 0
8	11,11,8	— 1/10 R	85 7 27	84 21 59	87 10 35
9	221	— 1/5 R	80 21 0	78 50 22	84 21 59
10	992	— 7/20 R	73 34 57	70 57 8	80 12 20
11	771	— 2/5 R	71 27 57	68 28 0	78 50 21
12	110	— 1/2 R	67 28 30	63 44 46	76 8 47
13	$88\bar{1}$	— 3/5 R	63 49 32	59 22 47	73 30 52
14	$55\bar{1}$	— 2/3 R	61 35 6	56 40 10	71 47 53
15	$33\bar{1}$	— 4/5 R	57 33 15	51 43 14	68 28 0
16	$55\bar{2}$	— 7/8 R	55 32 14	49 12 3	66 39 22
17	$22\bar{1}$	— R	52 32 30	45 23 26	63 44 46
18	$55\bar{3}$	— 8/7 R	49 37 5	41 34 34	60 35 25
19	$11,11,\bar{7}$	— 6/5 R	48 34 49	40 11 25	59 22 47
20	$33\bar{2}$	— 5/4 R	47 43 45	39 2 29	58 20 41
21	$77\bar{5}$	— 4/3 R	46 25 3	37 14 44	56 40 10
22	$44\bar{3}$	— 7/5 R	45 27 26	35 54 29	55 22 27
23	$55\bar{4}$	— 3/2 R	44 8 56	34 3 5	53 30 15
24	$66\bar{5}$	— 11/7 R	43 18 10	32 49 33	52 13 17
25	$77\bar{6}$	— 13/8 R	42 42 43	31 57 26	51 17 16
26	$11\bar{1}$	— 2 R	39 26 25	26 52 43	45 23 26
27	$13,13,\bar{14}$	— 9/4 R	37 52 12	24 15 13	42 1 18

No.	Symbol of Miller.	Symbol of Naumann.	1/2 (Apical Angle).	Inclination of face to Axis.	Inclination of Edge to Axis.
28	$77\bar{8}$	— 5/2 R	36° 37′ 31″	22° 4′ 20″	39° 2′ 29″
29	$55\bar{6}$	— 11/4 R	35 39 0	20 14 7	36 23 59
30	$33\bar{4}$	— 7/2 R	33 42 44	16 9 11	30 4 57
31	$55\bar{7}$	— 4 R	32 54 51	14 13 16	26 52 43
32	$11,11,\overline{16}$	— 9/2 R	32 20 45	12 41 43	24 15 13
33	$22\bar{3}$	— 5 R	31 55 24	11 27 40	22 4 20
34	$77,\overline{11}$	— 6 R	31 21 32	9 35 23	18 40 14
35	$33\bar{5}$	— 8 R	30 46 41	7 13 19	14 13 16
36	$44\bar{7}$	— 11 R	30 24 58	5 15 55	10 26 36
37	$55\bar{9}$	— 14 R	30 15 29	4 8 29	8 14 24
38	$11\bar{2}$	∞ R	30 0 0	0 0 0	0 0 0
39	$99,\overline{19}$	28 R	30 3 54	2 3 54	4 9 8
40	$17,17,\overline{37}$	18 R	30 9 24	3 13 24	6 25 35
41	$55,\overline{11}$	16 R	30 11 54	3 37 31	7 13 18
42	$44\bar{9}$	13 R	30 17 56	4 27 32	8 51 52
43	$33\bar{7}$	10 R	30 30 8	5 47 18	11 27 40
44	$22\bar{5}$	7 R	31 0 33	8 14 25	16 9 11
45	$55,\overline{13}$	6 R	31 21 32	9 35 23	18 40 14
46	$33\bar{8}$	11/2 R	31 36 19	10 26 36	20 14 7
47	$44,\overline{11}$	5 R	31 55 24	11 27 40	22 4 20
48	$11\bar{3}$	4 R	32 54 51	14 13 16	26 52 43
49	$22\bar{7}$	3 R	34 52 12	18 40 14	34 3 5
50	$11\bar{4}$	5/2 R	36 37 31	22 4 20	39 2 29

Skalenohedrons.

Table of Angles.

I. Zone ($\bar{h}01$).

No.	Symbol of Miller.	Symbol of Naumann.	$^1/_2$ (Angle of X-Edge).	$^1/_2$ (Angle of Y-Edge).	$^1/_2$ (Angle of Z-Edge.)	Inclination of X-Edge to Axis.	Inclination of Y-Edge to Axis.	Inclination of Z-Edge to Axis.
			° ′ ″	° ′ ″	° ′ ″	° ′ ″	° ′ ″	° ′ ″
	110	$- ^1/_2$ R	67 28 30	90 0 0	22 31 30	76 8 47	63 44 46	63 44 46
51	320	$- ^1/_5$ R 3	72 21 58	81 16 19	27 1 35	68 28 0	„	84 21 59
52	530	$- ^1/_8$ R 5	73 36 34	79 9 25	28 3 16	66 39 22	„	86 28 19
53	210	$^2/_3$ P 2	75 40 24	75 40 24	29 39 47	63 44 46	„	90 0 0
54	730	$^1/_{10}$ R 7	72 57 41	72 18 18	30 50 56	„	61 31 4	87 10 35
55	11,40	$^1/_5$ R $^{11}/_3$	70 19 32	78 54 25	31 56 32	„	59 22 47	84 21 59
56	310	$^1/_4$ R 3	69 2 20	79 41 46	32 27 9	„	58 20 41	82 58 14
57	720	$^1/_3$ R $^7/_3$	66 56 35	80 59 13	33 15 2	„	56 40 10	80 39 41
58	410	$^2/_5$ R 2	65 18 41	81 59 49	33 50 33	„	55 22 27	78 50 22
59	920	$^5/_{11}$ R $^9/_5$	64 0 29	82 48 25	34 17 43	„	„	77 21 49
60	510	$^1/_2$ R $^5/_3$	62 56 35	83 28 5	34 39 5	„	53 30 16	76 8 47
61	610	$^4/_7$ R $^3/_2$	61 19 4	84 29 25	35 10 22	„	52 13 17	74 15 35
62	710	$^5/_8$ R $^7/_5$	60 6 59	85 14 13	35 32 23	„	51 17 16	72 52 2
63	810	$^2/_3$ R $^4/_3$	59 13 28	85 48 22	35 48 21	„	50 34 41	71 47 53
64	910	$^7/_{10}$ R $^9/_7$	58 29 40	86 15 16	36 13 58	„	50 1 14	70 57 8
65	14,10	$^4/_5$ R $^7/_6$	56 24 25	87 33 38	36 34 23	„	48 24 3	68 28 0
	100	R	52 32 30	90 0 0	37 27 30	„	45 23 26	63 44 46
66	$17,0\bar{2}$	R $^{19}/_{15}$	51 28 41	85 47 41	44 8 30	55 22 27	40 11 25	„
67	$70\bar{1}$	R $^4/_3$	51 17 49	84 52 30	45 36 40	53 30 15	39 2 29	„
68	$60\bar{1}$	R $^7/_5$	51 10 28	84 0 8	47 0 40	51 43 14	37 56 48	„
69	$50\bar{1}$	R $^3/_2$	51 2 54	82 46 35	48 58 21	49 12 3	36 23 59	„
70	$40\bar{1}$	R $^5/_3$	50 57 39	80 56 25	51 56 6	45 23 26	34 3 5	„
71	$70\bar{2}$	R $^9/_5$	50 58 28	79 38 9	54 3 14	42 39 46	32 21 27	„
72	$30\bar{1}$	R 2	51 5 28	77 54 54	56 52 18	39 2 29	30 4 57	„
73	$50\bar{2}$	R $^7/_3$	51 26 10	75 33 38	60 46 44	34 3 5	26 52 43	„
74	$20\bar{1}$	R 3	52 18 55	72 12 8	61 29 16	26 52 44	22 4 19	„
75	$17,0\bar{9}$	R $^{13}/_4$	52 38 43	71 15 48	68 7 11	24 51 50	20 39 59	„

No.	Symbol of Miller.	Symbol of Naumann.	$^1/_2$ (Angle of X-Edge).	$^1/_2$ (Angle of Y-Edge).	$^1/_2$ (Angle of Z-Edge).	Inclination of X-Edge to Axis.	Inclination of Y-Edge to Axis.	Inclination of Z-Edge to Axis.
			° ′ ″	° ′ ″	° ′ ″	° ′ ″	° ′ ″	° ′ ″
76	$70\bar{4}$	R $^{11}/_3$	53 9 53	69 57 57	70 24 22	22 4 20	18 40 14	63 44 46
77	$50\bar{3}$	R 4	53 32 47	69 6 52	71 55 58	20 14 7	17 19 25	„
78	$80\bar{5}$	R $^{13}/_3$	53 58 50	68 23 27	73 14 16	18 40 14	16 9 11	„
79	$30\bar{2}$	R 5	52 30 40	67 13 49	75 22 12	16 9 11	14 13 16	„
80	$19,0\overline{13}$	R $^{16}/_3$	54 46 44	66 45 31	76 14 55	15 7 37	13 24 56	„
81	$10,0\bar{7}$	R $^{17}/_3$	55 1 28	66 20 35	77 1 56	14 13 16	12 41 43	„
82	$10,0\bar{8}$	R $^{19}/_3$	55 27 23	65 44 39	78 21 18	12 41 43	11 27 40	„
83	$40\bar{3}$	R 7	55 49 24	65 4 58	79 26 18	11 27 40	10 26 36	„
84	$50\bar{4}$	R 9	56 38 47	63 54 22	81 44 54	8 51 52	8 14 24	„
85	$60\bar{5}$	R 11	57 12 13	63 10 2	83 14 0	7 13 18	6 48 4	„
86	$13,0\overline{11}$	R 12	57 25 8	62 53 36	83 47 33	6 36 31	6 15 15	„
87	$70\bar{6}$	R 13	57 36 14	62 39 45	84 16 0	6 5 27	5 47 18	„
88	$90\bar{8}$	R 17	58 8 28	62 0 58	85 36 35	4 38 11	4 27 32	„
89	$10\bar{1}$	∞ P 2	60 0 0	60 0 0	90 0 0	0 0 0	0 0 0	„

II. Zone $\left(\frac{h}{k}\ 1\ \bar{1}\right)$.

No.	Symbol of Miller.	Symbol of Naumann.	$^1/_2$ (Angle of X-Edge).	$^1/_2$ (Angle of Y-Edge).	$^1/_2$ (Angle of Z-Edge).	Inclination of X-Edge to Axis.	Inclination of Y-Edge to Axis.	Inclination of Z-Edge to Axis.
	100	R	52 32 30	90 0 0	37 27 30	63 44 46	45 23 26	63 44 46
90	$51\bar{1}$	$^2/_5$ R 3	61 47 9	76 19 37	45 9 54	51 43 14	45 23 26	78 50 22
91	$41\bar{1}$	$^1/_4$ R 5	64 7 30	73 5 9	46 39 51	49 12 3	„	82 58 14
92	$31\bar{1}$	$^4/_3$ P 2	67 56 49	67 56 49	48 43 11	45 23 26	„	0 0 0
93	$52\bar{2}$	— $^1/_5$ R 7	64 3 17	70 50 40	49 57 52	„	42 39 45	84 21 59
94	$73\bar{3}$	— $^2/_7$ R 5	62 28 1	72 3 3	50 23 33	„	41 34 24	81 58 43
95	$21\bar{1}$	— $^1/_2$ R 3	58 41 34	74 56 27	51 12 34	„	39 2 29	76 8 47
96	$95\bar{5}$	— $^2/_3$ R $^7/_3$	55 57 25	77 3 30	51 37 59	„	37 14 44	71 47 53
97	$32\bar{2}$	— R $^5/_3$	50 57 39	80 56 25	51 56 6	„	34 3 5	63 44 46
98	$75\bar{5}$	— $^8/_7$ R $^3/_2$	49 0 53	82 27 45	51 54 3	„	32 49 33	60 35 25
99	$43\bar{3}$	— $^5/_4$ R $^7/_5$	47 37 35	83 33 2	51 50 23	„	31 57 26	58 20 40
	$11\bar{1}$	— 2 R	39 26 25	90 0 0	50 33 35	„	26 52 43	45 23 26
100	$32\bar{3}$	— 2 R $^3/_2$	43 2 56	81 35 45	61 16 25	30 4 57	20 14 7	„
101	$53\bar{5}$	— 2 R $^5/_3$	44 9 17	79 40 2	63 44 39	26 52 43	18 40 14	„

No.	Symbol of Miller.	Symbol of Naumann.	1/2 (Angle of X-Edge).	1/2 (Angle of Y-Edge).	1/2 (Angle of Z-Edge.	Inclination of X-Edge to Axis.	Inclination of Y-Edge to Axis.	Inclination of Z-Edge to Axis.
			° ′ ″	° ′ ″	° ′ ″	° ′ ″	° ′ ″	° ′ ″
102	$21\bar{2}$	— 2 R 2	46 4 39	76 37 50	67 39 16	22 4 19	16 9 11	45 23 26
103	$31\bar{3}$	— 2 R 3	49 59 13	71 14 51	74 40 30	14 13 16	11 27 40	„
104	$41\bar{4}$	— 2 R 4	52 15 4	67 26 58	78 23 11	10 26 36	8 51 52	„
	$10\bar{1}$	∞ P 2	60 0 0	60 0 0	90 0 0	0 0 0	0 0 0	„

III. Zone $\left(\frac{h}{k}\ 1\ \bar{2}\right)$.

	100	R	52 32 30	90 0 0	37 27 30	63 44 46	45 23 26	63 44 46
105	$41\bar{2}$	4 R 3/2	40 4 51	81 11 53	66 39 37	16 9 11	10 26 36	26 52 43
106	$31\bar{2}$	— 1/2 R 5	57 17 16	68 52 53	64 15 14	30 4 57	26 52 44	„
107	$52\bar{4}$	— R 3	52 18 55	72 12 8	61 29 16	26 52 44	22 4 19	63 44 46
	$21\bar{2}$	— 2 R 2	46 4 39	76 37 50	67 39 16	22 4 19	16 9 11	45 23 26
108	$32\bar{4}$	— 5 R 7/5	38 27 16	82 30 8	66 0 38	14 13 16	8 51 52	22 4 20
109	$43\bar{6}$	— 8 R 5/4	35 39 38	84 49 14	64 31 23	10 26 36	6 5 27	14 13 16
	$11\bar{2}$	∞ R	60 0 0	90 0 0	60 0 0	0 0 0	0 0 0	0 0 0
	$01\bar{2}$	R 3	52 18 55	72 12 8	61 29 16	26 52 44	22 4 19	63 44 46
	$\bar{2}1\bar{2}$	— R	52 32 30	90 0 0	37 27 30	63 44 46	45 23 26	„

IV. Forms not in the foregoing Zones.

$\frac{h-1}{h+k+1}$	Symbol of Miller.	Symbol of Naumann.	1/2 (Angle of X-Edge).	1/2 (Angle of Y-Edge).	1/2 (Angle of Z-Edge.	Inclination of X-Edge to Axis.	Inclination of Y-Edge to Axis.	Inclination of Z-Edge to Axis.
			° ′ ″	° ′ ″	° ′ ″	° ′ ″	° ′ ″	° ′ ″
1.11	$11,4\bar{3}$	7/6 P 2	69 19 56	69 19 56	44 55 18	49 12 13	49 12 13	0 0 0
1.455	$10,7\bar{6}$	— 10/11 R 8/5	52 27 33	81 54 58	48 35 21	49 34 15	41 33 37	65 50 56
1.8	$16,10,\bar{11}$	— R 9/5	50 58 28	79 38 9	54 3 14	42 39 46	32 21 27	63 44 46
1.846	$15,7\bar{9}$	— 8/13 R 3	56 14 18	73 52 1	56 28 16	36 28 39	33 22 54	73 6 56

$\frac{h-l}{h+k+l}$	Symbol of Miller.	Symbol of Naumann.	½ (Angle of X-Edge).	½ (Angle of Y-Edge).	½ (Angle of Z-Edge.)	Inclination of X-Edge to Axis.	Inclination of Y-Edge to Axis.	Inclination of Z-Edge to Axis.
			° ′ ″	° ′ ″	° ′ ″	° ′ ″	° ′ ″	° ′ ″
1.888	32,14,$\overline{19}$	—5/9 R 17/5	57 1 6	72 43 36	57 16 39	38 25 44	33 5 30	74 40 34
,,	12,2$\overline{5}$	1/3 R 17/3	60 7 28	69 35 32	57 52 13	34 3 5	37 14 46	80 39 49
1.905	25,11,$\overline{15}$	—4/7 R 10/3	56 46 48	72 50 33	57 26 36	38 15 15	32 49 34	74 15 35
1.951	50,21,$\overline{30}$	-22/41 R 40/11	57 13 33	72 4 21	58 7 7	37 19 53	32 23 49	75 10 35
2	52$\overline{3}$	— 1/2 R 4	57 38 16	71 16 1	58 55 5	36 23 59	31 57 26	76 8 47
,,	63$\overline{4}$	— 4/5 R 5/2	53 40 0	75 17 25	57 49 20	37 56 48	30 48 29	68 28 0
,,	71$\overline{3}$	2/5 R 5	58 59 3	69 54 31	59 10 53	35 54 29	32 21 29	79 50 22
2,16	91$\overline{4}$	1/2 R 13/3	57 28 12	70 21 41	60 54 24	34 3 5	30 4 57	76 8 47
2.20	64$\overline{5}$	—7/5 R 11/7	46 36 6	81 13 3	57 6 48	37 56 48	26 52 43	55 22 27
2.25	534	— 5/4 R 9/5	48 25 16	79 4 12	58 34 4	36 23 59	26 52 43	58 20 41
2.285	95$\overline{7}$	— 8/7 R 2	49 43 3	77 33 15	58 33 2	35 21 35	26 52 43	60 35 25
2.33	42$\overline{3}$	— R 7/3	51 26 10	75 33 38	60 46 44	34 3 5	26 52 43	63 44 46
2.40	73$\overline{5}$	— 4/5 R 3	53 48 55	72 49 47	62 19 51	32 21 27	26 52 43	68 28 0
2.591	10,5$\overline{8}$	— 8/7 R 9/4	50 14 22	73 12 35	62 24 38	31 40 46	24 35 56	60 35 25
2.66	51$\overline{3}$	8/3 P 2	62 45 10	62 45 10	66 18 2	26 52 44	26 52 44	0 0 0
2.857	131$\overline{7}$	4/7 R 5	56 29 27	68 24 17	66 56 33	26 52 43	23 55 3	74 15 35
3.35	35,17,$\overline{32}$	-31/21 R 67/31	47 58 31	75 46 8	66 15 31	25 30 12	19 16 5	53 56 30
3.385	23,11,$\overline{21}$	-20/13 R 11/5	48 9 27	75 30 47	66 31 57	25 12 16	19 7 36	52 48 30
3.5	41$\overline{3}$	— 1/2 R 7	57 17 51	66 5 46	70 59 33	22 4 20	20 14 7	76 8 47
3.88	17,10,$\overline{18}$	— 7/3 R 5/3	43 21 24	79 31 37	65 20 57	23 28 58	·16 9 10	40 59 16
4	71$\overline{5}$	4 P 2	61 19 22	61 19 22	73 41 19	18 40 14	18 40 14	0 0 0
4.33	72$\overline{6}$	— R 13/3	53 58 50	68 23 27	73 14 16	18 40 14	16 9 11	63 44 46
4.5	51$\overline{4}$	— 1/2 R 9	57 32 44	64 34 36	75 0 3	17 19 25	16 9 11	76 8 47
4.591	21,3,$\overline{11}$	16/7 R 2	45 35 45	76 30 44	68 53 59	19 32 6	14 13 16	41 34 24
4.66	81$\overline{6}$	14/3 P 2	60 59 21	60 59 21	75 55 7	16 9 12	16 9 12	0 0 0
4.80	15,$\overline{19}$	8/5 R 3	50 27 26	71 26 17	72 44 32	17 34 39	14 13 16	51 43 14
5	82$\overline{7}$	— R 5	52 30 40	67 13 49	75 22 12	16 9 11	14 13 16	63 44 46
,,	11,2$\overline{9}$	— 1/2 R 10	57 40 40	64 3 28	76 26 36	15 39 25	14 39 37	76 8 47
,,	13,$\overline{27}$	5/2 R 2	45 20 15	76 26 51	69 36 40	18 0 4	13 2 46	39 2 29
5.33	11,3$\overline{5}$	4 R 4/3	38 4 48	83 32 36	64 6 25	18 40 14	11 27 40	26 52 43
,,	91$\overline{7}$	16/3 P 2	60 45 58	60 45 58	77 37 16	14 13 16	14 13 16	0 0 0

$\frac{h-l}{h+k+l}$	Symbol of Miller.	Symbol of Naumann.	½ (Angle of X-Edge).	½ (Angle of Y-Edge).	½ (Angle of Z-Edge).	Inclination of X-Edge to Axis.	Inclination of Y-Edge to Axis.	Inclination of Z-Edge to Axis.
			° ′ ″	° ′ ″	° ′ ″	° ′ ″	° ′ ″	° ′ ″
5.75	12,3,$\overline{11}$	$-\frac{5}{4}R\frac{28}{5}$	53 42 1	67 37 50	76 33 12	14 13 16	12 21 30	58 20 41
6	11,$\bar{1}\bar{7}$	2 R 3	49 59 13	71 14 51	74 40 30	14 13 16	11 27 40	45 23 26
„	10,$\bar{1}\bar{8}$	6 P 2	60 36 37	60 36 37	78 57 30	12 41 43	12 41 43	0 0 0
6.5	71$\bar{6}$	$-\frac{1}{2}R\,13$	58 1 58	63 0 44	79 29 32	12 2 49	11 27 40	76 8 47
6.66	95,$\bar{1}\bar{1}$	$-4R\frac{5}{3}$	41 46 40	79 15 19	68 46 31	14 13 16	9 35 24	26 52 43
6.87	35,$\bar{7}$,$\overline{20}$	$\frac{29}{8}R\frac{55}{29}$	43 51 31	77 7 14	70 46 16	13 24 57	9 55 53	29 13 5
8	5$\bar{1}\bar{3}$	4 R 2	44 28 22	76 14 24	72 3 59	11 27 40	8 14 25	26 52 43
„	13,1,$\overline{11}$	8 P 2	60 20 52½	60 20 52½	81 40 32	9 35 23	9 35 23	0 0 0
12	7$\bar{1}\bar{5}$	4 R 3	49 19 56	70 59 2	77 49 28	7 13 18	5 47 18	26 52 43
∞	7$\overline{25}$	∞ R 2	76 6 6	76 6 6	73 53 54	0 0 0	0 0 0	0 0 0
∞	5$\overline{14}$	∞ R 3	70 53 36	70 53 36	79 6 24	0 0 0	0 0 0	0 0 0
1.86	65,30,$\overline{39}$	$-\frac{17}{28}R\frac{52}{17}$	56 20 25	73 40 17	56 39 23	39 14 35	33 16 34	73 19 44

The following forms will be found in other authors, but on account of uncertainty in their determinations and other assigned reasons have been omitted.

From the zone $\left(\frac{h}{k} 11\right)$.

— $^1/_4$ R — According to Zippe, in Hausmann, but without designation of the combination.

$^3/_{10}$ R — Determined by Zippe from measurements of Bournon.

— $^9/_8$ R — Hessenberg VII: from Lake Superior. This form is — $^8/_7$ R. See Appendix I.

— $^{10}/_7$ R — Determined by Websky in a comb. from Striegau in Silesia.

$^{10}/_{13}$ R — H. III. 267. On crystals of dolomite from the Binnenthal.

— $^7/_4$ R — H. VII: from Agaete.

— $^{12}/_5$ R — A curved face determined by Zippe without measurements,

$^{13}/_4$ R — Zippe after Bournon.

9 R — H. VI: from Iceland. He afterwards substituted 10 R for this.

16 R — H. IV: 6 comb. from Bleiberg.

25 R — H. X: from Andreasberg.

$^7/_4$ R — According to Zippe in Dufrenoy, but given in no comb. It is also in Bournon.

$^9/_2$ R — Zippe determined this form from measurements of Bournon. He gives it as uncertain.

From the zone $\left(\frac{h}{k} 0 \bar{1}\right)$.

Schnorr gives the 2 following forms $^{11}/_{17}$ R $^{10}/_{11}$ and $^5/_8$ R $^{20}/_{13}$ the latter of which, however, does not quite fall in the zone. Herr Schnorr was good enough to inform me, that on repeating his measurements he could get no very satisfactory results; and he also sent some of the original crystals to the Bonn Museum, where I had an opportunity of examining them. I convinced myself that no accurate determinations could be made upon them.

R $^7/_4$ — according to Zippe is given by Hausmann but not in a combination.

R 15 — is given in Zippe's figg. 68. 69, but in neither case could it be accurately measured.

From the zone $\left(\frac{h}{k}\ 1\bar{1}\right)$.

— $^7/_5$ R $^9/_7$ — Zippe took this form from Dufrenoy. See his remarks on it. Websky also gives it on calcite from Striegau in Silesia. The faces were curved however and admitted of no exact measurement.

— $^5/_3$ R $^{17}/_{15}$ — Zippe, fig. 15, substitutes this form for. $e^{3/2}$ in Levy's fig. 122. It seems to me that this figure must be entirely rejected because the faces could not possibly have the positions indicated in it. The edge $e_{3/2} : e^1$ is necessarily in the edge-zone of e^1. The edge $i : e_{3/2}$ is drawn parallel to $e_{3/2} : e^1$ with all possible accuracy. The form i is thus in this zone and the edge i:i should be parallel to the other edges. This is not the case in the drawing. The face i is given as $(d^{1/9}\, d^{1/5}\, b^{1/6}) = -\,^7/_8$ R $^{15}/_7$. Zippe replaces this by — $^5/_4$ R $^{13}/_9$.

— 2 R $^5/_2$ — Levy, var. 195. See the remarks to Zippe's fig. 32.

Faces not in the above zones.

$(d^{1/10}\, d^{1/21}\, b^{1/19})$ — Levy, var. 83. There are no contradictions in the determination of this form, but, as Weiss and Zippe have remarked, we are entirely without any means of checking it, either from the geometry of the combination or from measured angles.

$(d^1\, d^{5/7}\, b^{2/5})$ — Levy, var. 169. See the remarks to Zippe's, fig. 75.

($b^{1/5}\, d^{1/2}\, d^{1/8}$) — Levy, var. 177. The face was curved and the determination inexact.

$-{}^{3}/_{2} R\, {}^{7}/_{3}$, $-{}^{1}/_{2} R\, {}^{7}/_{3}$, $-{}^{7}/_{4} R\, {}^{25}/_{21}$, $-{}^{1}/_{2} R\, {}^{13}/_{3}$, $-{}^{1}/_{4} R\, {}^{71}/_{3}$, $3 R\, {}^{25}/_{9}$, $10 R\, {}^{17}/_{5}$, ${}^{17}/_{2} R\, {}^{23}/_{17}$, ${}^{11}/_{2} R\, {}^{19}/_{11}$, will found in Zippe's work, he having taken them from Bournon. For this last form Descloizeaux substitutes ($b^{1/11}\, d^{1/3}\, d^{1/7}$), $10 R\, {}^{9}/_{5}$.

${}^{154}/_{3} R\, {}^{35}/_{35}$ — Zippe, fig. 49, determined this form from a zone and one angle both of which were observed with a goniometer without telescop (I infer this from the remaks to fig. 67). The faces of the form were in general curved, only a small number being sufficiently plain for measurement. It is so possible that even these, when examined with a goniometer with two telescopes and a light would show irregularities which would effect both the measurement and the observed zone, that the form must be considered as doubtful. Descloizeaux substitutes for this ($d^{10/11}\, d^{1}\, b^{1/2}$).

$-161 R\, {}^{25}/_{23}$ — Zippe, fig. 67. This face was also determined by a zone and an angle. Of the zone I dont think there can be any doubt although one of the faces ∞R was perfectly dull. Of the angle however there is. The faces of the angle are extremely small and do not cut each other. This rendered it necessary in measuring that the eye should be placed some distance from the crystal. Thereby the chances of error were increased. In addition to this, the angle chosen for measurement, the Y-angle, is the one which on such steep forms varies the least, for a given variation of the indices. I think therefore the form is doubtful.

$2 R\, {}^{11}/_{4}$ & ${}^{32}/_{35} R\, {}^{11}/_{4}$ are two skalenohedrons determined by Hessenberg (III) on crystals from Rossie, St. Laurence Co. N.-Y. In Miller's symbols they are respectively $(41\,\bar{4}\,\overline{25})$ & $(61\,\bar{1}\,\overline{27})$. Calculations and measurements were too widely different, however, to establish forms with such high indices. $(10\bar{1}\bar{6})$, $(2R\,^8/_3)$ and $(110\bar{5})$, $(R\,^8/_3)$ agree better with Hessenberg's measurements than the given forms. I have omitted them from the tables, however, because I think the combination needs reinvestigation.

$-\,^3/_5 R\,3$, $-\,^{17}/_{15} R\,^{37}/_{17}$, $-\,^7/_8 R\,^{27}/_{11}$, $^{14}/_{27} R\,^{11}/_3$, $-\,^{22}/_{25} R\,^{13}/_8$ $^1/_3 R\,^{23}/_{25}$ belong to 2 combinations from Lake Superior, described by Hessenberg in VII, which I have not been able on the reinvestigation to confirm. The same is true of $-\,^1/_2 R\,^7/_2$, $^{10}/_{21} R\,4$, $^1/_5 R\,^{19}/_3$, $^{10}/_{17} R\,^{35}/_9$, forms determined on crystals from Agaete, Gran Canaria (H VII), and of $-\,^5/_4 R\,^{17}/_{11}$ in a combination from Andreasberg (H VIII). See Appendix I.

$-\,^1/_3 R\,15$ & $-\,8 R\,5$ — have been determined by the same author in 2 combinations from Andreasberg. They are however thoroughly uncertain. The three angles, X, Y, Z, of $-\,^1/_3 R\,15$ were measured. From these taken two and two, three values of the indices may be calculated and by comparing these among themselves we may determine whether the measurements agree with one another or not. Instead however of comparing the values of the indices we may compare the values of n in m R n which are much more easily calculated. The numbers obtained were 9, 11, $12^1/_3$, which are very widely apart. With regard to $-$ 8/R 5 Hessenberg himself was in doubt. $-\,7 R\,5$ agrees as well with the observations as $-\,8 R\,5$.

I regret to say, that I think the following forms determined by my honored teacher, Prof. G. vom Rath of Bonn are doubtful. In each case they are faces which on measurement gave varying angles. From the great constancy of the angle of the cleavage form of calcite, in its pure varieties, I do not think, that a variation of anything like a half of a degree will be found in the angles of well developed faces. Where such variations occur I think they will most always be found to be due to roundings composed of several faces. It is especially easy to understand how such roundings could be produced in the case of faces requiring very special condition for their development. A slight change of the conditions, as, for example, of the rate of evaporation of the solvent, or of the quantity of the solution, would cause a new face to appear. If the change were sufficiently gradual, we should expect to find the two limiting faces connected by a series of intermediate faces, which would approach more or less to a continuously curved surface. It is obvious that in such a case, the problem is to find the two extreme faces and as many of the intermediate ones as possible. It is not to refer the rounding to a single face. On the majority of minerals such determinations are not necessary. On calcite however they are. Its ability to assume the most varied crystallographical forms, makes it peculiarly sensitive to changes of the surrounding conditions, while in the act of crystallisation.

For the investigation of these cases a goniometer with two telescopes and a light is necessary. When a brillant object as a flame is reflected from such a broken surface each part of it gives a separate image and from them the true nature of the apparent curvature may be ascertained. I am sure it would have been quite imposible to decipher the combinations described in appendix I by any other means.

In all of the determinations which I think are doubtful, Prof. vom Rath has made the measurements in the ordinary way with the image of a bar in the window frame for an index. I have had very little experience with measuring crystals in that way, but I feel very certain it is much less satisfactory than with the goniometers of Fuess, for example. I venture therefore to suggest, that these crystals may probably yield other results if investigated in the manner above indicated.

— $^{36}/_{35}$ R 2 — R 19 d, from Andreasberg. The faces of this form had a decided curvature. Descloizeaux substituts for it — R 2, ($d^{1/7}$ $d^{1/4}$ $b^{1/5}$).

$^6/_7$ R 3 — R 5 n. F. Bergen Hill. This form is only suggested; not given as certainly determined.

— $^6/_5$ R $^5/_3$ — R 5, n. F. Bergen Hill. In addition to the angles X, Y, Z, those to — $^7/_5$ R a. + R were also measured. From the last two, I attempted to calculate the form. I could, however, arrive at no satisfactory result.

— $^{11}/_{12}$ R $^7/_3$ — R 5, n. F. Bergen Hill. With this form occur two others given as — $^{19}/_{15}$ R $^{91}/_{57}$ & — $^{17}/_{12}$ R $^{73}/_{51}$ & — 2 R. Of the angles measured on these forms certain of them are incompatible with each other.

For — $^{19}/_{15}$ R $^{91}/_{57}$.

Y = 161° 50′; 58′; 162° 2′; 16′. Angle to — 2 R = 171° 3′.

For — $^{17}/_{12}$ R $^{73}/_{51}$.

Y = 166° 10′; 2′; 2′; 165° 56′; 54′; 50′; 44′; 42; 42; 40° 38′.

Angle to — 2 R = 172° 50′.

It is easy to see that the particular rhombohedron which truncates the edge of a skalenohedron makes a greater angle with the faces at that edge than any other rhombohedrom. According to the measurements, such a rhombohedron should make an angle with the faces at the Y-Edge of
— $^{19}/_{15}$ R $^{91}/_{57}$ = 170° 55′; 59′; 171° 1; 8′; of
— $^{17}/_{12}$ R $^{73}/_{51}$ = 173° 5′; 1′; 1′; 172° 58′; 57′; 52′; 51′; 51′; 50′; 49′.

— 2 R does not truncate the edge and consequently must make a smaller angle with the faces than the truncating rhombohedron. According to the extreme measurements, however, in both of the above instances the reverse of this is the case. The observed difference is quite small, but I think it sufficient to show, that further investigation of these crystals is necessary.

$^1/_3$ R 23 — $(38,2,\overline{31})$. Sella, Quadro. No. 287. Unaccompanied by figure or measurements. Descloizeaux substitutes $(b^{1/17}\ d^1\ d^{1/21})$.

$(d^{1/21}\ d^{1/15}\ b^{1/29})$ or $(d^{1/10}\ d^{1/14}\ b^{1/9})$. Descloizeaux, fig. 268. One symbol is given in the text, the other in the figure. The faces of the form were a little curved.

($d^{1/6}\ d_1\ b^{1/5}$). Descloizeaux substitutes this form for ($d^{1/39}\ d^{1/5}\ b^{1/32}$) which he gives as Levy's. Levy, however, gives no such form.

— $^{5}/_{3}$ R $^{17}/_{15}$ — Dana fig. 560. Rossie: given as doubtful.

— $^{3}/_{2}$ R $^{13}/_{9}$ — A curved face determined by Bauer (Min. Mit. Zeitschrift d. d. geol. Gesellschaft. 1874 p. 397.)

APPENDIX I.

The following are the results of a reinvestigation of certain crystals described by Hessenberg[1]).

1. **From Agaëte, Grand Canary.** (H VII.)

Fig. 1 represents a crystal corresponding to Hessenbergs fig. 4. It is all but a true picture of the crystal, magnified twice. Some of the details have been varied in position so as to bring more into the figure.

The principal surfaces of the crystal are striations parallel to edge of primitive. On the summit of the crystal this approaches a plane which would be — $^1/_2$ R (110). On the sides it is between the faces R 3, and is curved. This latter Hessenberg has given as ∞ P 2. In some places it really approachs R 5, as is seen by the lines of intersection with 4 R, $(3\bar{1}\bar{1})$ becoming parallel.

4 R, $(3\bar{1}\bar{1})$, R 3, $(20\bar{1})$ and R, (100) are of great perfection as regards planeness and lustre. The rhombohedrons — $^2/_3$ R, $(55\bar{1})$ and — $^4/_3$ R, $(77\bar{5})$ have been correctly determined by Hessenberg. The faces of the former vary very much in size; sometimes they encroach upon — $^1/_2$ R to the extent indicated in Hessenberg's figure. I observed this, however, in but one case. The faces of — $^4/_3$ R are always small in this combination. 0 R is also present.

As will be seen from my figure, the faces designated by Hessenberg as $^{10}/_{21}$ R 4 and $^1/_5$ R $^{19}/_3$ are apparently in a zone with R 3 and — $^1/_2$ R. These faces are usually rounded and partially dull. Their number is either 2 or 3. The edges between them are not very sharp. At one point, however, a well developed

1) In this appendix I have used principally Naumanns symbols in order that a comparison with Hessenberg may be easier.

face was found; and from that I was enabled to establish the zone. The face is $^1/_3$ R $^{17}/_3$, (12 2 $\bar{5}$).

$^1/_3$ R $^{17}/_3$ V R = 21° 55′ observed 21° 55′

„ V R 3 (of zone) = 11° 41′ „ 11° 32′.

In addition to these the following inclinations were observed on less perfect faces: to R 21° 50′; to R 3, 11° 21′; 11° 28′; 11° 20′. The images of this face were always more or less extended into a band of light, lying wholly in the zone. One portion was, however, always much more prominent than the rest.

A face always appeared betwen this face and — $^1/_2$ R. This was entirely dull and of varying curvature. Sometimes it was approximately plane, with its edge truncated by + R. If this were really the case it would be $^4/_3$ P 2. By adjusting the telescope upon it and observing when it appeared to be brightest, I was enabled to measure approximately its inclination to R 3. The angle is of course the mean of a number of such observations.

Observed 21° 53′ $^4/_3$ P 2 V R 3 22° 17′.

Below $^1/_3$ R $^{17}/_3$ was a face, which was sometimes quite separated from it, and at other times was a mere rounding. When in the better developed condition its inclination to R 3 was different on different specimens. Measured approximately in the above indicated manner the inclinations to R 3 were

3° 43′	. . .	$^8/_{11}$ R $^7/_2$, (19, 1$\bar{9}$)	V R 3	3° 38′
5 27	. . .	$^5/_8$ R $^9/_5$, (13, 1$\bar{6}$)	V R 3	5 26
9 40	. . .	$^2/_5$ R 2, (7, 1$\bar{3}$)	V R 3	9 58.

I do not consider these determinations, with the exception of that of $^1/_3$ R $^{17}/_3$, sufficiently good to admit the forms to the table.

Several crystals, of which Hessenberg's fig. 5 is the type, are represented by my figures 2, 3, 4, 5. They consist of the following zones and faces, combined in different ways on each particular crystal.

The rhombohedral zone. Hessenberg gives the following forms, all of which I can most accurately comfirm. 0 R, R, 4 R, ∞ R, — 11 R, — $^{13}/_8$ R, — $^4/_3$ R, — $^2/_3$ R. In measuring one of the zones I observed an image corresponding to — $^3/_2$ R. The edge between — $^4/_3$ R and — $^2/_3$ R is rounded. This is represented in my figure 5 as if it were a separate face.

The edge-zone of 4 R. Hessenberg observed the following forms, which I can most exactly comfirm. 4 R 2, 4 R $^3/_2$, 4 R $^4/_3$, R 3. These forms are not always lettered in my figures, but they will be easily understood, if it is remembered, that, when only two of the first three skalenohedrons are present, the second is the missing one.

The zone 4 R 2 : R 3 (just above as seen in centre of fig. 5.) This zone was observed by Hessenberg but his symbol of the skalenohedron, ($^{10}/_{17}$ R $^{35}/_9$), does not satisfy it. It is identical with the zone R 3 : — $^1/_2$ R, of fig. 1 and forms a connection between the two combinations, so different in appearance. The face in this zone is curved and lustrous. On the best specimens its image was wholly in the zone. On others it was principally in the zone, but with more or less diffuse reflexion outside of it. This curvature is about limited by two faces of the zone, $^1/_2$ R $^{13}/_3$ and $^2/_5$ R 5. Sometimes the one sometimes the other was more prominent. In either case a band of light proceeded from the image of the one in the direction of that of the other. Sometimes both images were present, sometimes the band of light gradually vanished at one end. The following measurements were made.

From R 3, 7° 12'; 7° 34'; 7° 35$^1/_2$', 9° 52'. In one case the band of light begun at 7° 3', was maximum at 7° 43' and faded away beyond 9°. The calculated angles are

$^7/_{13}$ R $^{29}/_7$; (20 2 $\bar{9}$) V R 3 = 7° 2'
$^{10}/_{19}$ R $^{21}/_5$; (29 3 $\overline{13}$) V R 3 = 7° 17'
$^1/_2$ R $^{13}/_3$; (9 1 $\bar{4}$) V R 3 = 7° 48$^1/_2$'
$^2/_5$ R 5 (7 1 $\bar{3}$) V R 3 = 9° 58'.

It will be seen from several of the above measurements, that the surface sometimes springs from a face nearer to R 3 than $^1/_2$ R $^{13}/_3$. I have only introduced the last two faces into the table. The second may perhaps be considered as established, as the image of 7° 12' quite marked.

$^1/_2$ R $^{13}/_3$ V R (cleavage) = 81° 7' observed 81° 16'
" V — $^4/_3$ R = 32° 40' " 32° 28'.

I have always given $^1/_2$ R $^{13}/_3$ in the figures as it is more frequent than $^2/_5$ R 5. A comparison of figures 2 and 4 will show the variations of size of $^1/_2$ R $^{13}/_3$.

Sometimes + R was long and narrow; and there appeared to be a zone $^{1}/_{2}$ R $^{13}/_{3}$: R: — $^{2}/_{3}$ R. Hessenberg gave this as one of the zones. When adjusted on the goniometer one very faint and separated image was really in the zone. Its distance from + R was 24° 7, 24° 3. In the zone are

$^{3}/_{5}$ R $^{11}/_{3}$; ($23\ 2\overline{10}$) V + R = 24° 18′
$^{14}/_{23}$ R $^{25}/_{7}$; ($35\ 3\overline{15}$) V + R = 23° 59′.

The face common to the two zones is $^{4}/_{7}$ R 4 ($11\ 1\ \bar{5}$). Its inclination to + R is 25° 23′. In one case a second image in the zone was observed at 25° 22′.

The small face in fig. 2 bounded by $^{1}/_{2}$ R $^{13}/_{3}$, R, — $^{2}/_{3}$ R was dull, but to all appearence in the zone R 3: $^{1}/_{2}$ R $^{13}/_{3}$. It was also curved.

There is a second zone determined by faces of R 3 and 4 R 2 on the crystals. The darker striation of fig. 2 and that of fig. 5 lie in it. There were no intermediate faces between R 3 and 4 R 2 which were well developed. The image from the striation was, however, a band of light. Some of the more distinct points at which it was crossed were, from R 3,

5° 27′ . . $^{11}/_{14}$ R $^{53}/_{11}$; ($33\ 1\overline{20}$) V R 3 = 5° 35′
6° 26′ . . $^{3}/_{4}$ R $^{47}/_{9}$; ($29\ 1\overline{18}$) V R 3 = 6° 26′
6° 53′ . . $^{8}/_{11}$ R $^{11}/_{2}$; ($27\ 1\overline{17}$) V R 3 = 6° 52′.

The following inclinations are those, to R 3, of a faint image, somewhat out of the zone. They were measured without disturbing the adjustment of the zone.

13° 18′; 13° 13′; 13° 23′.

They correspond to $^{16}/_{3}$ P 2. Its calculated inclination to R 3 = 13° 30′. When the images were properly adjusted this was observed to be 13° 28′. The angle between two such images corresponding to the apical of $^{16}/_{3}$ P 2 was observed to be 121° 35; calculated 121° 32′. The same on well developed faces was observed to be 121° 33. A number of angles of this form to various others were measured with perfectly satisfactory results.

The original of fig. 5 was the most prominent example of this zone. The striae between R 3 and 4 R 2 arranged themselves principally in two planes. One of these I have designated as $^{16}/_{3}$ P 2. It was quite impossible, however, to connect them with any of the separate images of the band of light.

This form, $^{16}/_{3}$ P 2, determines with R 3 another, as it may be called, undeveloped zone, which is important for the combinations figg. 2, 3, 4. In fig. 3 a striation will be seen connecting the two faces. In figg. 2, 4 the striation limiting the body of the crystals is in this zone. It is but little different in position from the last zone described. Sometimes the striae lie partly in one zone partly in the other. If so they are slightly curved and their images consist of two slighly divergent bands of light. From figure 4 it will be seen that the striation approaches a form which would cut $^{1}/_{2}$ R $^{13}/_{3}$ in a horizontal edge; this would be $^{4}/_{5}$ R $^{13}/_{3}$, $(33\ 1\ \overline{19})$. The most prominent images of the band were in general quite near $^{16}/_{3}$ P 2. From this face, I observed images at $54^{1}/_{2}'$; $1^{0}\,2^{1}/_{2}'$; $1^{0}\,3'$; $1^{0}\,6'$; $1^{0}\,8^{1}/_{2}'$; $1^{0}\,9'$; $1^{0}\,18'$; $6^{0}\,36'$; $6^{0}\,54^{1}/_{2}'$; $6^{0}\,56'$; $7^{0}\,12'$; $12^{0}\,36'$; $13^{0}\,27^{1}/_{2}'$; $13^{0}\,28'$.

Calculated inclinations.

$^{1}/_{10}$ R 51;	$(29\ 3\ \overline{22})$	V $^{16}/_{3}$ P 2	=	$0^{0}\ 51'$
$^{2}/_{17}$ R 43;	$(49\ 5\ \overline{37})$	V „	=	1 2
$^{1}/_{7}$ R 35;	$(20\ 2\ \overline{15})$	V „	=	1 13
$^{4}/_{5}$ R $^{13}/_{3}$;	$(33\ 1\ \overline{19})$	V „	=	8 33 (no corresponding image observed)
R 3	. . . .	V „	=	$13^{0}\ 30'$.

On running over the zones of the figures they will be found to be the rhombohedral zone and zones of which R 3 is one of the determining faces.

I also examined the crystal corresponding to Hessenberg fig. 2. The correctness of the determination of $^{1}/_{10}$ R 7 I can confirm. The face — $^{1}/_{2}$ R $^{7}/_{2}$ is, however, utterly indeterminable.

2. **From Lake Superior.**

Fig. 6 represents one side of the crystal corresponding to Hessenberg's fig. 6. The other side may be described as follows. The edge of δ on the extreme right bounds a bright plane face which breaks into steps with δ. This passes backwards to — $^{8}/_{7}$ R. The faces β and γ are not present. Adjacent to — $^{8}/_{7}$ R is + R. Between this face + R and the one on the left of the figure is a Y-Edge of — $^{4}/_{5}$ R 3. This is surmounted by small faces, ε, β, β_1.

A symmetrical figure, as usually given, would convey no correct idea of this crystal.

The faces + R, 4 R and 0 R were, of course, correctly determined. His — $^9/_8$ R, on the contrary, was not. This face is — $^8/_7$ R (55$\bar{3}$).

— $^8/_7$ R VR (opposite)	= 93° 2′	observed	93° 3$^1/_2$
			93 3
— $^8/_7$ R VR (either side)	= 42° 42′	„	42 42
			42 41

This form determines with + R a zone in which almost all the faces on the crystal lie. Since + R is in the zone all of them must have two indices with the common ratio 5 : 3.

— $^4/_5$ R 3; (7$\bar{3}$5) is the form upon whose extremity all the others are placed.

— $^4/_5$ R 3 V + R (of zone)	= 38° 10′;	observed	38° 15$^1/_2$′
			38 14
			38 23
Y-Edge of — $^4/_5$ R 3	= 145° 42′;	observed	145 39 [1]).

This is the form given by Hessenberg as — $^7/_8$ R $^{22}/_{11}$. There are in reality a number of forms in this one, all lying near together. — $^4/_5$ R 3 is easily selected, however, as the most prominent. The face which intersects + R on the left of the figure gave an image which was but little blurred. This not the rule however.

β = — $^8/_{13}$ R 3; (5 $^7/_8$ $\bar{3}$)

β V + R (of zone)	60° 44′;	observed	60° 51′
β V + R (adjacent, not of zone)	34° 12′;	„	34 10.

This face is striated. It is not separated with perfect sharpness from γ.

γ = — $^4/_7$ R $^{10}/_3$; (5 $\overline{^{11}/_3}$ $\bar{3}$).

γ V + R (of the zone)	61° 53′	observed	61° 56$^1/_2$′
γ V + R (on right of figure)	33° 54′	„	33 52$^1/_2$
γ $V\beta$	1° 9′	„	1 5$^1/_2$.

---

1) As, to a certain extent, in conformity with the tables, I have given the direct angles of the edges X, Y, Z of the skalenohedrons. The remaining angles are those between the normals to the faces.

The images from β and γ were connected by a band of light, which was crossed at two points. These were $12^1/_2$ and 41 minutes of angle distant from β. The following two forms correspond to them.

$$-\,{}^{17}/_{28}\,R\,{}^{52}/_{17};\ ({}^1/_6,\ {}^1/_{13},\ \overline{{}^1/_{10}})\ V\beta = 13'.$$
$$-\,{}^{10}/_{17}\,R\,{}^{16}/_5;\ (5\ {}^9/_4\,\overline{3})\ V = 43'.$$

The latter face has not been introduced into the table. The former having been otherwise observed has been.

The observed inclination of β to + R should probably be two or three minutes smaller than that given. The proper image of β was several minutes broad (perhaps 4). I think, I might with correctness have chosen the beginning, instead of the middle as the proper point, from which to measure; because this partial blending of the images of the several faces would indicate a progression from one to the other rather than an oscillation of one about the other. This would bring the observed and calculated results several minutes nearer together. If β were alone present one might be in doubt whether it were $-\,{}^8/_{13}$ R 3 or $-\,{}^{17}/_{28}$ R ${}^{52}/_{17}$. Since, however, an image corresponding to the latter was also observed all doubt is removed.

$\alpha = -\,{}^{10}/_{11}\,R\,{}^8/_5,\ (5\ {}^7/_2\ \overline{3})$.

$\alpha\ V+$ R (of zone)	$51^0\ 43'$	observed	$51^0\ 52'$
$\alpha\ V+$ R (to left, above)	$37^0\ 45'$	„	$37\ \ 37^1/_2$.

This face is striated parallel to axis of zone.

$\beta_1 = -\,{}^{17}/_{28}\,R\,{}^{52}/_{17},\ ({}^1/_6,\ {}^1/_{18},\ \overline{{}^1/_{10}})$, the face above mentioned, 13, from $-\,{}^8/_{13}$ R 3.

$\beta_1\ V+$ R (of zone)	$60^0\ 57'$ . . .	observed	$60^0\ 55^1/_2'$
$\beta_1\ V+$ R (above)	33 9 . . .	„	33 7.

$\varepsilon = -\,{}^5/_9\,R\,{}^{17}/_5,\ (32\ 14\ \overline{19})$.

$\varepsilon\ V+$ R (to right)	$61^0\ 52'$ . . .	observed	$61^0\ 56'$
			61 55
			61 54
$\varepsilon\ V+$ R (to left)	$33^0\ 37'$ . . .	observed	33 37
			33 37.

This face I at first took for γ. Each time, however, I observed a slight deviation of the image from the zone. This with constancy of the latter angle constrained me to think differently. The same inclination to + R (to left) 33° 37′ was observed on the crystal correspönding to Hessenberg fig. 3. Its separation from β is not perfectly sharp on the crystal, but the images were entirely unconnected by diffuse light, which was not the case with β β_1 and γ.

As above mentioned small faces corresponding to β β_1 and ε are found on the other side of the crystal surmounting the Y-Edge of — $^4/_5$ R 3. On one side of the edge there were two faces, on the other side only one. The double faces have the following inclinations to 2 faces, + R.

$$\left\{ \begin{matrix} 60^0\,50' \\ 34\quad 9 \end{matrix} \right\} \therefore = \beta. \qquad \left\{ \begin{matrix} 61^0\,56' \\ 33\quad 37 \end{matrix} \right\} \therefore = \varepsilon.$$

ε is unmistakeable. β is, however, not so plain. It might from these observations alone be β or β_1. The observations are however almost identical wich those by which β was estabtished above when β_1 was also present.

The single face opposite thesc and apparently identical with one of them seems however to be β_1. That it is either β or β_1 is certain. Its inclinatiou to β above ascertained would indicate the latter. The calculated angle is 32° 28′ observed 32° 30′. If it were β, this would be its X-Edge = 32° 16′.

From the face behind and those in front the following angles could also be measured.

X-Edge ε	114° 2′	observed	114° 3′
$\beta : \beta_1$ (over + R)	112 35	„	112 39½
$\beta_1 : \varepsilon$ (on rear of crystal)	146 23	„	146 21.

The face mentioned above as forming the edge with δ on the right is also in the zone. It is

— $^{22}/_{41}$ R $^{40}/_{11}$; $(5021\overline{30})$: or $(^5/_3, {}^7/_{10}, \overline{1})$.

Inclination to	+ R (of zone)	62° 45′;	observed	62° 41′
„ „	+ R (above δ)	33 39;	„	33 34
„ „	γ (across δ)	65 59½;	„	65 59.

Although this face breaks into steps, one of its parts was quite plane and lustrous and gave a good image, which was accompanied by a faint band of light in the direction of — R. This began with an faint image 40′ distant from — $^{22}/_{41}$ R $^{40}/_{41}$, corresponding to $- {}^{4}/_{7}$ R ${}^{10}/_{3}$. This band of light, I think, establishes the fact, that the face is in the zone + R, $- {}^{8}/_{7}$ R. The measurement to the other face + R confirms this. I am not sure if this face be not $(45\ 19\ \overline{27})$ which has exactly the observed inclination to + R of zone. The latter could not, however, like the former, be expressed in a second, it may be simpler, manner. I mean its indices could not be written in a form corresponding to $({}^{5}/_{3}, {}^{7}/_{10}, \bar{1})$. The face $- {}^{1}/_{2}$ R 4, according to vom Rath so characteristic of calcite from this locality, would be in this zone and only about 1° of angle from $(50\ 21\ \overline{30})$ in the direction from + R. No trace of an image corresponding to it was observed.

θ is a striation parallel to the edge $- {}^{22}/_{41}$ R ${}^{40}/_{11}$: δ and corresponding to the steps of this face on the rear of the crystal. This I ascertained in the following manner. The reflexion from these striae as seen in the goniometer with Websky's slit consist of two bands perpendicular to the striae. When the zone + R: $- {}^{8}/_{9}$ R : β : γ was adjusted one could turn the instrument until the centre of the cross-hairs occupied a point midway between the two bands of light which passed obliquely through the field of view. The angle read off corresponds to a face in the zones + R: $- {}^{8}/_{7}$ R and that of the striae. The angle, counted from + R, was 62° 36′. The face is evidently $- {}^{22}/_{41}$ R ${}^{40}/_{11}$ for which 62° 41′ was observed. The edge θ: $- {}^{4}/_{5}$ R 3 was apparently vertical whem the zone $- {}^{4}/_{5}$ R 3: + R: $- {}^{8}/_{7}$ R was adjusted.

δ is quite plane but dull. I measured approximatly its inclination to + R and its Y-Edge, by reading off the angle at which the reflected light seemed to reach its maximum. The results were.

$\delta\ V + R = 20^\circ\ 42'$, mean of 8 readings between 20° 14′ and 21° 14′.

$\delta\ V - {}^{22}/_{41}\ R\ {}^{40}/_{11} = 15^\circ\ 31'$, mean of 3 readings between 15° 18′ and 15° 48′.

$\delta\ V \gamma = 16^\circ$.

δ, Y-Edge = 145° 20′, difference of means of 7 readings for each face; the extremes of the one being 1° 17′ apart, of the other 1° 1′.

The form approximates to ${}^{1}/_{2}$ R ${}^{11}/_{8}$, $(81\bar{3})$; for which inclination to + R = 20° 55′;

to — $^{22}/_{41}$ R $^{40}/_{11}$ = 15° 27;
to γ = 15° 56′;
Y-Edge = 144° 37′.

The Y-Edge of δ is probably truncated by an extremely narrow face of a rhombohedron. When the zone 4 R: R (above) was adjusted, this edge shone quite brightly in one position appoximately 10° 57′ from R. The Y-Edge of $^1/_2$ R $^{11}/_3$ makes an angle of 11° 20$^1/_2$′ with + R (above).

δ being know, it was attempted to determine θ from the two zones above mentioned, but no form with simple indices could be found. It approximates to — $^1/_2$ R $^{13}/_3$ (83$\bar{5}$).

The second combination described by Hessenberg, from Lake Superior, is his fig. 3. I have not redrawn this. His figure will answer, if we imagine certain changes. R an 0 R are correct. I could not, however, find a trace of — $^9/_8$ R. The crystal has only one free end. The attached end is bounded by the lower skalenohedron, which expands considerably. The appearance of the crystal is, therefore, pyramidal. The edge between the upper and lower skalenohedron is not sharp, but rounded.

The upper skalenohedron. Its faces are easily seen, with the naked eye, to be composed of a number of forms all lying very close together. As appeared from the reflected image these are principally two and they correspond to forms of the other combination. They are — $^5/_9$ R $^{17}/_5$ and — $^{17}/_{28}$ R $^{52}/_{17}$

— $^5/_9$ R $^{17}/_5$; (32 14 $\overline{19}$).

X-Edge	= 114° 2′	. . .	observed	114° 4′
Y- „	= 145 25	. . .	„	145 25
				145 14.

— $^{17}/_{28}$ R $^{52}/_{17}$; (65 30 $\overline{39}$); ($^1/_6$, $^1/_{18}$, $\overline{^1/_{10}}$).

X-Edge	= 112° 41′	. . .	observed	112° 35′
Y- „	= 147 20$^1/_2$	. . .	„	147 30
				147 20$^1/_2$.

It is scarcely to be doubted, that images, corresponding to the other forms of the first crystal, lying near these two, could be recognised by sufficiently careful observation. It seems as if the confusion of the images were only apparent and that they could be analysed into accurately measurable parts.

The lower skalenohedron would by analogy be expected to be the same as the bearer of the first combination. In spite of its extremely broken and striated surface I obtained one measurement of the Y-Edge which confirms this. On one face I had no doubt of the image which should be chosen. On the other there were two corresponding to the angles 145° 57; 145° 45′. The Y-Edge of — $^4/_5$ R 3 is calculated to 145° 42′.

The following is a list of all the forms observed in the principal zone of the combinations.

$(5\ \bar{7}\ \bar{3})$, — $^4/_5$ R 3
$(1\ 1\ 0)$, + R
$(5\ 5\ \bar{3})$, — $^8/_7$ R
$(5\ ^7/_2\ \bar{3})$, — $^{10}/_{11}$ R $^8/_5$
$(5\ ^7/_3\ \bar{3})$, — $^8/_{13}$ R 3
$(5\ ^{30}/_{13}\ \bar{3})$, — $^{17}/_{28}$ R $^{52}/_{17}$
$(5\ ^9/_4\ \bar{3})$, — $^{10}/_{17}$ R $^{16}/_5$
$(5\ ^{11}/_5\ \bar{3})$, — $^4/_7$ R $^{10}/_3$
$(5\ ^{21}/_{10}\ \bar{3})$, — $^{22}/_{41}$ R $^{40}/_{11}$.

The observed form not in the zone were 0 R, (111); 4 R, $(31\bar{1})$; $^1/_2$ R $^{11}/_3$, $(81\bar{3})$; $^5/_9$ R $^{17}/_5$, $(32\ 14\ \overline{19})$.

3. From Andreasberg.

The crystals of the combination — $^5/_4$ R, — $^5/_4$ R $^{17}/_{11}$. I have also examined (H VIII). The rhombohedron I did not measure. The skalenohedron I did. Its faces were finely pimpled. In same cases they were, to the naked eye, apparently plane. This was discovered, however, not to be the case as soon as they were examined by reflexion. I attempted to determine the form by measuring its inclinations to the three cleavage planes. This I did 3 times.

I.	II.	III.
138° 54′	139° 15′	139° 12′
119 20	119 56	120 39
106 4	106 26 . .	105 31[1]).

1) Direct angles, if one imagines the face as truncating the lateral corner of the primitive.

The first and third measurements of the second set were quite sharp, but I could deduce no satisfactory symbol from them. The intercepts of the face exactly corresponding to them are $1 : 0.5602 : \overline{0.7370}$. For 0.5602 one may set $^{9}/_{16} = 0.5625$ and for 0.7370, $^{17}/_{23} = 0.739$. These intercepts, expressed in whole numbers, would be $368 : 207 : \overline{272}$.

A face to which the general surface of the pimples nearly approaches is $(16\ 9\ \overline{12})$; $(1 : 0.562 : 0.075)$; $- {}^{14}/_{13}\ R\ 2$.

Its inclinations to the cleavage planes are

$$139^{\circ}\ 34'$$
$$120\quad 9$$
$$105\quad 57.$$

From the above measurements, however, one could by no means consider it a certainly determined face.

There are three other combinations of Hessenberg, which I think need reexamination. They are those, from Rossie N.-Y., with $2\ R\ ^{11}/_{4}$, $^{32}/_{35}\ R\ ^{11}/_{4}$ (III), from Andreasberg with $-8\ R\ 5$ and $-25\ R$ (X) and from Andreasberg with $- {}^{1}/_{3}\ R\ 15$ and $^{1}/_{5}\ R\ ^{11}/_{8}$. They are not in the collection now belonging to the university of Halle and therefore they did not come into my hands with those I have described. Remarks on them will be found in the list of doubtful forms.

I think the question raised by vom Rath and Hessenberg as to the validity of the law of simple intercepts receives no support from these crystals. Some of the forms I have given as corresponding to images in band of light have not as simple indices as could be wished. They will be found, however, to have the simplest indices of any form about the required position. I hope soon to explain how they have been calculated.

As examples of zonal development these crystal possess a certain uniqueness; especially that first described from Lake Superior.

Exception having been taken to my assigning separate symbols to the faces γ and ε on the crystal fig. 6 from Lake Superior, I have thought it well to explain at some length the way in which one should proceed in cases where such fine distinctions are to be drawn. The problem is precisely analogous to the corresponding one of astronomy: the observations must be discussed and weighed.

The objection raised was, of course, that the observed differences between the angles of the two faces fall within limits, between which angles have been observed to vary.

The variations of the angles of a crystal depend first upon the nature of its substance, some crystals having very constant angles and others not. We may easily infer that those crystals whose angles vary most, are most easily disturbed in the process of growth. The second element upon which variations depend is the quantity and quality of the disturbances, which are different in each particular case. Variations of angle are thus specific for each crystal but they are much more probable for some crystals than others. The problem is therefore to recognise when variations occur and as far as possible to eliminate them.

A certain general notion of the trueness with which a crystal has been formed may be gathered from the condition of the more common faces (evenness and lustre) and their inclinations to one another. The results must be applied with caution, however, to other faces than those actually observed.

For a particular face the first thing to be observed is, whether it gives a perfectly sharp image, as a truly polished surface, and if not, how far it approximates to such. Faces which give a single, sharply defined image can not be regarded as otherwise than regularly built. It may well be that in the process of growth they have departed from the theoretical inclinations to distant parts of the crystal with which they have only a mediate connection, but one can not suppose that to be true as regards the particles immediately in them. The practical rule is, therefore, when possible, always to measure to the nearest faces and to those portions of them immediately adjacent to the given face.

This is illustrated by the following. Faces on opposite ends of a crystal of any size, which should be parallel, are very often not so. Faces of a zone which extend all around a crystal are very often not accurately in the zone though they theoretically should be. From a cleavage face of calcite some lines in extent images as much as a degree apart may often be obtained, although sufficiently small rhombohedrons show variations of only one or two minutes.

By measuring faces near together, the irregularities of growth between faces more removed from one another are eliminated.

I think it will often be found, in measuring from single but somewhat broad images, that one or the other of their extremities and not their middle point corresponds to faces with the simplest intercepts, which are probably the faces of most stable growth.

By measuring in the above manner one is enabled to reduce the probable errors of the observations, and to ascertain more certainly, where two approach very nearly, whether they belong to one and the same form. There are often accidents of faces which enable one to do this still more certainly. Whenever fine distinctions are to be drawn the nature of the particular case must be intelligently considered.

We will now review the reasons for which we give to γ and ε separate symbols. The observed inclinations of γ to $+$ R (of zone) and of ε to $+$ R (to right) may be called identical. The calculated angles differ by only 1′. The observed inclinations to the other $+$ R faces are for ε identical with the calculated and for γ $1^1/_2$′ different, although they are for the two faces $15^1/_2$′ different, an error scarcely to be expected where only one well defined image was present. The images from them were also different. The image of γ was connected to that of β_1 and β by a band of light which seems to me to indicate very certainly that the zone is real. β_1 and ε were connected by no such band; they were entirely separated. γ was exactly in the zone with R and $- {}^8/_7$ R; ε deviated from it. If the accidents in the two cases were the same instead of different one would be inclined to think the observed differences of angle due to irregularities of growth. This not being the case they confirm the difference of the faces indicated by the angles.

The angles of the crystal seem generally to be quite constant. The differences between observed and calculated angles are generally under 5′. The angles just mentioned are measured from the nearest faces which were well developed. The angles of the faces β, β_1, ε, γ to each other agree quite well with the calculations. So also do the remaining angles above given. It will be found that a substitution of γ for ε, or β for β_1 will introduce much larger differences between observed and calculated results for all the angles. The faces adjacent to β_1 and ε are not the same as those adjacent to β and γ. The above circumstances convinced me that the faces γ and ε belong to different forms, though I did not appreciate them at once, Accordingly I have assigned different symbols to them.

The distinction of β in one case and β_1 in another rests only on measured angles. These, however, can be much more certainly trusted after the observed differences for γ and ε have been confirmed by other circumstances.

BONN, PRINTED BY CHARLES GEORGI.

Calcite from Agaëte and Lake Superior.

1.

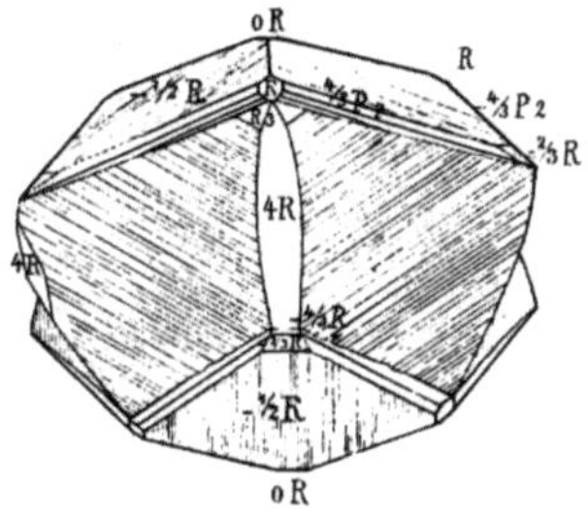

2.

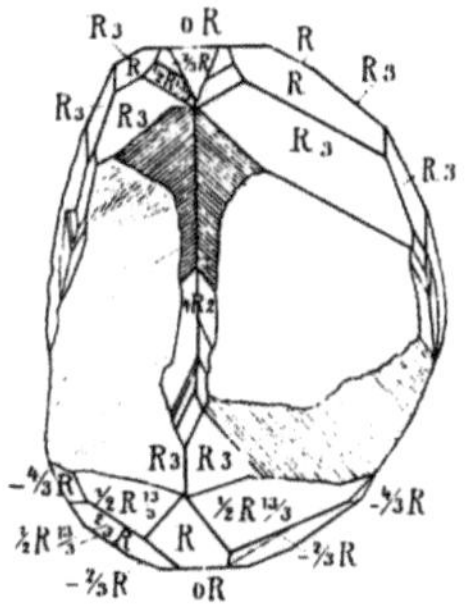

3.

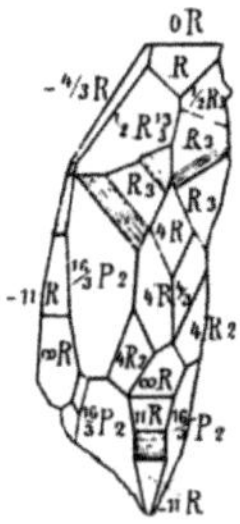

4.

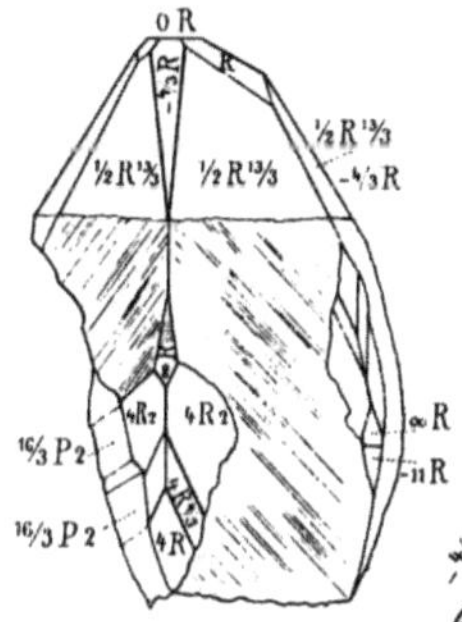

5.

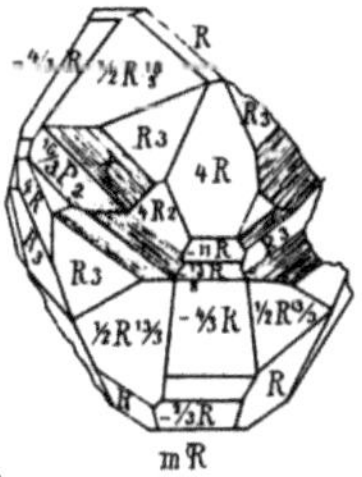

6.

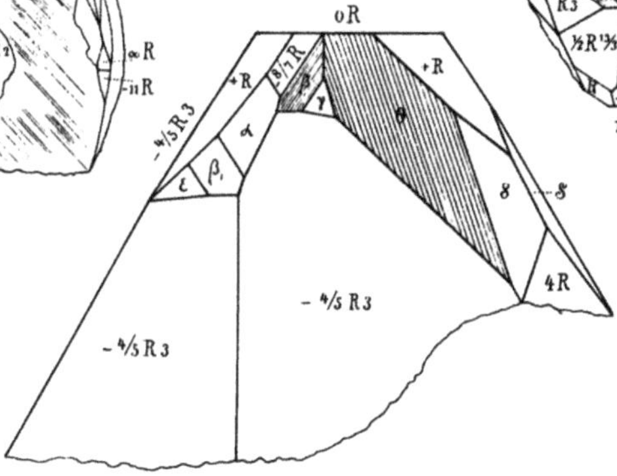

J. Jrby del.

Lith.Jnst v.A.Henry, in Bo

Bei Fragen zur Produktsicherheit wenden Sie sich bitte an:
If you have any questions regarding product safety,
please contact:

Walter de Gruyter GmbH
Genthiner Straße 13
10785 Berlin
productsafety@degruyterbrill.com